Certainty: A Refutation of Scepticism

Certainty:

A Refutation

of Scepticism

Peter D. Klein

UNIVERSITY OF MINNESOTA PRESS
Minneapolis

Publication of this book was assisted by a grant
from the publications program of the National Endowment
for the Humanities, an independent federal agency.

Published by the University of Minnesota Press,
2037 University Avenue Southeast, Minneapolis, MN 55414
Printed in the United States of America

Library of Congress Cataloging in Publication Data

Klein, Peter D. (Peter David), 1940 -
    Certainty, a refutation of scepticism.
    Includes index.
    1. Certainty 2. Skepticism. I. Title.
BD171.K55   121'.63   81-13040
ISBN 0-8166-0995-0   AACR2
ISBN 0-8166-0998-5 (pbk.)

To my teachers:

Mr. Hale, my teacher in the sixth grade;
Wayne Booth and Grimsley Hobbs, my teachers in college;
Rulon Wells, my teacher in graduate school;
Roderick Chisholm, my teacher in books.

Without them, I could not have made
even the errors in this book, much less have
possibly arrived at anything right.

CONTENTS

PREFACE

When I first studied Descartes' evil genius argument in an introductory philosophy class, I thought that the argument just could not be correct. There had to be a good refutation of it. But Descartes' own refutation, depending as it did upon a purported demonstration of the existence of an epistemically benevolent god, seemed far less persuasive than the considerations which originally had led him to delineate the class of "things that can be doubted." And other purported refutations, which I studied later, appeared less convincing than the arguments for scepticism.

Twenty years later, my students react to Descartes' argument about the "evil spirit, not less clever and deceitful than powerful" in the same way that I did. They, too, believe that there must be something wrong with the argument for scepticism—but what is it? That is what this book is about.

I realize that it is presumptuous to believe that I have found an answer; in fact, the evil genius argument, at times, still seems sound to me. Whereas Descartes employed the evil genius hypothesis to help him resist the power of his "former opinions" until he had satisfied himself that some propositions were known with absolute certainty, I have just the opposite difficulty. The evil genius argument, perhaps updated somewhat by substituting the brain-in-the-vat for the evil spirit, is still so entrancing that I have to rehearse my objections to it over and over again in order to balance my former acceptance of it. The evil spirit is clever, indeed!

There is one, important cautionary note I wish to issue: This is a book in the history of philosophical scepticism. "Scepticism" has meant many things, and I will be discussing only three varieties. I think they are the ones that many contemporary philosophers would not find most plausible. But it may be that Descartes did not really have any one of these varieties precisely in mind. It would take a Cartesian scholar to make such a claim. Consequently, this book should not be viewed as a discussion of particular historical sceptical positions, including the one that I call *Pyrrhonian Scepticism*; rather, what you have before you is an examination of what I hope you believe are among the *most* plausible versions of scepticism.

Nevertheless, there are other candidates for the most plausible versions of scepticism and other ways to attempt refutations. In fact, several books that approached scepticism from a different perspective were published during the time that I was working on my book.* I began to read one of them and realized that I would be unable to finish this book without first taking account of what the other authors had written. Since there was no good reason to believe that the series of new books on scepticism would ever come to an end, I decided that the only way to finish was to postpone studying them. I have now read them; and I was right! Had I read them earlier, I would not now be writing this preface.

Although it is difficult to resist commenting on these books, I dare not impose any further on the patience of the editors of Minnesota Press—perhaps the following short sketch of my book will serve to distinguish it from those of other recent critics of scepticism.

This book has four chapters. Chapter One contains a brief description of the three forms of scepticism I seek to refute and an outline of the strategy for pursuing that goal, including a discussion of two important constraints on the debate between the sceptics and nonsceptics. The first is that the sceptic cannot begin by insisting that evidence, in order to be adequate for empirical knowledge, deductively imply the purportedly known proposition. If that were the

---

*Stanley Cavell, *The Claim of Reason*, Oxford University Press (Clarendon Press, Oxford) 1979; James W. Cornman, *Skepticism, Justification and Explanation*, D. Reidel Publishing Company, Dordrecht, Holland, 1980; and Oliver A. Johnson, *Skepticism and Cognitivism*, University of California Press, Berkeley, 1978.

starting point, there could be no debate. For scepticism concerning the possibility of empirical knowledge would follow immediately. On the other hand, I believe that we must grant to the sceptic that empirical propositions, if known, are absolutely certain. To fail to grant that would be to deny a fundamental intuition motivating scepticism. Consequently, one way of putting the thesis of this book is: Absolute certainty is possible on the basis of non-deductive evidence. In fact, I will argue that empirical, contingent propositions are no less evidentially certain than tautologies!

The task of Chapter Two is to identify and then evaluate the epistemic maxims and principles upon which scepticism is based. I defend one sceptical epistemic principle against various attacks and show that it can be derived from a general model of justification which is designed to be acceptable to sceptics and non-sceptics as well as to foundationalists and coherentists. The model is used to assess three other sceptical epistemic principles. Then the arguments for scepticism employing the four epistemic principles are examined. Needless to say, I argue that no one of them provides a good reason for scepticism.

Chapter Three is primarily a construction and defense of an account of knowledge which employs the model of justification developed in Chapter Two. The account is then used to show how a proposition can be absolutely certain on the basis of non-deductive inferential evidence. In addition, I use the account of knowledge to offer a solution to the so-called "lottery paradox" which has been thought to provide solace for the sceptic.

Chapter Four utilizes the results of Chapters Two and Three to show that the varieties of scepticism identified in Chapter One are implausible.

I have many people to thank for helping me with this book. My colleagues at Rutgers University were supportive in many ways. They were always willing to listen to my "latest" thought about some particular point; they often forced me to develop an even later one. Special thanks are due Martha Bolton, Martin Bunzl, Mary Gibson, Richard Henson, Howard McGary, Frederic Schick, LaVerne Shelton, Arthur Smullyan, and a graduate student of mine, Al Shepis. Each of them read various versions of the manuscript, made telling criticisms of it, but were often kind enough to help me solve the problems they had identified.

Several people outside of Rutgers read various drafts of the manuscript and provided me with excellent comments: Robert Audi, John Barker, Roderick Chisholm, Ed Erwin, George Pappas, Frederick Schmidt, David Shatz, Michael Smith, and Peter Unger. They all helped me to avoid some errors; but no doubt, others remain which are solely my responsibility.

Loretta Mandel typed the manuscript at each stage of its development from a forty-page paper to its present size. She probably believes that it is her book — and she is partly correct. Shelby Netterville, a graduate student of mine, helped me in preparing the index. The Rutgers Research Council helped to pay for some of the typing expense.

Finally, I want to thank the staff at the University of Minnesota Press.

I hope the book proves to be worth all the effort that others have put into it.

Certainty: A Refutation of Scepticism

CHAPTER 1

HOW TO REFUTE SCEPTICISM

1.1 The Goal and Strategy

In the *Treatise*, Hume wrote that sceptical doubts "arise naturally" from careful philosophical investigation and claimed, albeit ironically, that "carelessness and in-attention alone can afford us any remedy."[1] I do not believe that he was correct; and the goal of this book is to show that scepticism is implausible by carefully examining the epistemic principles upon which it is based.

Aside from the philosophically unsatisfactory "remedy" for scepticism mentioned by Hume, there are two initially attractive methods often employed in attempting to demonstrate the implausibility of scepticism. The first is to reject scepticism merely because it conflicts with the supposedly obvious claim that we do have knowledge; the second is to defend an analysis of knowledge which contains either a relatively weak set of necessary and sufficient conditions of knowledge or a set specifically designed to be immune to sceptical attacks.

However, there is a third strategy not usually employed. It is to meet the sceptic on his/her own ground by granting as much as possible while at the same time showing why what can be granted does not lead to the sceptical conclusion. I believe that only the third strategy can successfully show that scepticism is implausible.

The first strategy begins by asserting that it is clear, regardless of anything else, that we do know many things which the sceptic claims that we cannot know. Just as we should reject any theory of motion which has the consequence that the swift Achilles cannot pass the slower turtle, we should discard any theory of knowledge which has the consequence that one cannot know those propositions which, beyond doubt, we do know.

Thus, as Moore[2] and, more recently, Chisholm have claimed, "The general reply to a scepticism that addresses itself to an entire area of knowledge can only be this: we do have the knowledge in question, and, therefore, any philosophical theory implying that we do not is false."[3]

Now, all that may be well and good, but many of the sceptical arguments designed to show that there can be no knowledge have proceeded from the very assumptions apparently embraced by the nonsceptic. Thus, the sceptic can point out that those who reject scepticism solely because it leads to the conclusion that there can be no knowledge are thereby rejecting the very epistemic principles which provide a foundation for their purportedly nonsceptical theories of knowledge.

In addition, the sceptical arguments seem convincing once one attends to them; and if the arguments remain plausible after careful scrutiny, at some point the so-called "theories of knowledge" will have to be understood as theories which lead to the counterintuitive conclusion that there can be no knowledge. Just as our well-entrenched intuitions about simultaneity seem to require revision in the light of new physical theories, so it may be that intuitions concerning the scope of knowledge may require revision once the epistemic principles are carefully investigated. To repeat, *the challenge of scepticism is that it appears to proceed from universally accepted epistemic principles.*

A second strategy designed to demonstrate the implausibility of scepticism is to develop an analysis of knowledge which contains conditions immune to the traditional sceptical attack. For example, if knowledge is true, reliable belief, and if it were proposed that the reliability of a belief depends solely upon its causal ancestry, scepticism could be rather easily discarded. It would merely require showing that a suitable, requisite causal history *could* obtain. For if those conditions could obtain, then knowledge is possible for us. But I take it that the sceptics' claim is not merely that no one in fact has (or is likely to have) knowledge. Their claim is stronger than that. It is that no one *could* have knowledge because at least one of the necessary conditions of knowledge *cannot* be fulfilled. The power of scepticism depends upon the initial plausibility of its assumptions, shared by both the sceptic and his/her critics, and upon the apparently *inescapable* conclusion that no one has knowledge. The claim is not merely that it is a fact that no one has knowledge. Rather, that is a fact because no one *can* have knowledge.

Furthermore, the sceptical tradition did not arise within a context in which the reliability of knowledge was believed to be dependent solely upon the causal ancestry of beliefs. That may have been an important aspect of the reliability requirement; but the context in which scepticism traditionally arose was one in which a belief is considered to be reliable only if it is adequately justified. Some, usually sceptics, have insisted that adequate justification obtains only when the justification provides for certainty. Thus, a "method" of refuting scepticism based upon an analysis of knowledge which explicates the reliability condition solely in terms of causal ancestry of beliefs ignores the epistemological tradition that gave rise to, and still provides the foundation for, scepticism.

As mentioned above, I propose to meet the sceptic on his/her own grounds and, as far as possible, to let him/her choose the weapons. In particular, I shall grant that knowledge, even inferential knowledge, is true, justified, and *absolutely certain* belief and I shall attempt to show that the epistemic principles underlying scepticism are either unacceptable or fail to provide adequate grounds for scepticism. Specifically, my strategy will consist in showing that the sceptics have failed to provide a cogent argument *for* scepticism and that there is a characterization of certainty which is, or at least ought to be, acceptable to the sceptic and which is such that many of our beliefs are certain. This is not unlike showing which of the seemingly plausible assertions leading to Zeno's paradoxical account of motion misleads us and how we might better capture the pretheoretical intuitions inaccurately portrayed by that misleading assertion.

1.2 Varieties of Scepticism

But, first, we must clarify what is meant by *scepticism*. For there are important differences between the varieties of scepticism, and arguments for one variety are not necessarily arguments for another. Consider, first, what I will call *Direct Scepticism.* It claims that no person, S, *can* know that p, where 'p' stands for any proposition ordinarily believed to be knowable.* I say "can" because their claim, as mentioned above, is not that S does not know that p (although that follows from their claim), but rather that some necessary condition of knowledge cannot be fulfilled. Thus, for example, their claim would be that even if S were to acquire more evidence for p, S would still not know that p because S can never acquire enough evidence, it would be alleged, in order to acquire knowledge that p. The strategy of Direct Scepticism is to locate at least one necessary condition of knowledge which cannot be fulfilled.

The *Iterative Sceptic* is not quite so ambitious. The claim here is that it is not the case that S can know that S knows that p. Some may believe that Iterative Scepticism is a form of scepticism with a rather short history beginning with the development of contemporary forms of epistemic logic. On the contrary, although sceptical literature may not always have distinguished Iterative from Direct Scepticism, I believe it will become apparent that often those arguments which were designed to support Direct Scepticism, if sound, would support only Iterative Scepticism. In fact, it is the conflation of those two forms of scepticism which provides a basis for some of the initial plausibility of the sceptical arguments.

The difference between Direct Scepticism and Iterative Scepticism is that only the Iterative Sceptic is willing to grant that it is possible for S to know that p, but he/she continues, even if S were to know that p, S cannot know that he/she knows that p. For example, suppose that in order for S to know (inferentially) that p, the inference used by S to arrive at p must have some property, Q (e.g., validity). Iterative Scepticism would assert that S cannot determine whether the inference has Q. What good is knowledge, the Iterative Sceptic might ask, if we cannot in practice distinguish it from mere belief?

The distinction between Iterative and Direct Scepticism is crucial if we are to understand the possible arguments for scepticism. It may be thought that if it cannot be *shown* that the principle of inference has Q, S cannot know that p. But all that would thereby be shown is that we cannot know whether we know that p. To put it another way, a necessary condition of knowledge that p may be that the belief that p must be sufficiently, perhaps absolutely, certain. As I

---

*The abbreviations 'p' and 'S', and others used throughout the book, are defined in sections 2.3 and 2.7.

said above, I am willing to grant that. Thus, if we could not show that a belief is certain, we could not show that the belief is certifiable as knowledge. However, we need not be able to *show* that a belief is certain in order for it to *be* certain. We need not be able to *show* that a principle of inference is valid in order for it to be valid. Finally, we need not be able to *show* that S can know that p, in order for S to know that p.

However, there is a feature of our speaking (Grice would call it "conversational implication"[4]) which may lead to conflating the considerations which would support Direct Scepticism with those which support Iterative Scepticism. Failing to distinguish those considerations leads to the further conflation of Direct with Iterative Scepticism. The feature of our speaking which I have in mind here can be illustrated in the following, perhaps typical, conversation between S and C. Such conversations may appear to provide evidence for the claim that S is justified in believing that p on the basis of some evidence, e, only if S can show that he/she is so justified.

S: Jones owns a Ford.
C: What justifies you in asserting that Jones owns a Ford?
S: Well, she says that she owns one, she drives one, and she has a valid-looking title to a Ford. That is what justifies my belief that Jones owns a Ford.
C: So now you are claiming that the evidence which you cited justifies the assertion that Jones owns a Ford. What justifies you in asserting that *that* evidence justifies you in asserting that Jones owns a Ford? . . . (etc.)

The sceptic might point out that in order to answer the question, "What justifies you in asserting that Jones owns a Ford?," S would, at least implicitly, assert something of the form "e justifies p for S." Since S is *now* claiming that e justifies p, S is obligated (in some sense) to defend that claim if challenged, and the only plausible way to do that is by providing evidence for the claim that e justifies p. Thus, if S claims ⌜p⌝,[5] S is obligated to provide a justification for p on the basis of some evidence, e, and that in turn obligates S to justify the assertion that e justifies p.

This seems to impose an unending set of obligations to defend assertions, which does seem contrary to our ordinary practices. As Chisholm and Wittgenstein have pointed out, some assertions seem to provide stopping points.[6] However, for the sake of the argument,

I am willing to grant that the chain of obligations is unending.

But what this shows, if it shows anything, is that if S *asserts* ⌜p⌝, S is obligated to justify the assertion by citing the evidence e, and *that* assertion further obligates S to justify the assertion that e justifies p. It does not provide any reason for believing that if e justifies S in believing that p, then S is justified in believing that e justifies p. It is the act of asserting ⌜p⌝ which places S in the position of having to provide a defense of p, if challenged, and that in turn, places S in the position of having to defend the assertion that e justifies p. But since a person may be *justified* in believing that p without asserting ⌜p⌝, or even, for that matter, believing that p, being *justified* in believing that p does not impose a similar chain of obligations. Thus, while *asserting* or *claiming* ⌜p⌝ may lead to an obligation to defend the claim that e justifies p, being justified in believing that p does not lead to an obligation to justify the belief that e justifies p.

Thus, I am willing to grant that asserting ⌜p⌝ may place S in the position of having to defend the claim and the way in which he/she arrived at it. S may not be able to do that, but S may nevertheless be justified in believing that p. For S is justified in believing that p on the basis of the evidence available to him/her for p. That is, S is justified in believing that p, if S has the epistemic right to believe that p; and S may have that right but be unable for a variety of reasons to produce an adequate defense of it.

An analogous situation may help to clarify the distinction here. Suppose that S has the right to a piece of property, P, as a result of some valid transfer, all traces of which have been lost. S may not know that he/she has that right to P; in fact, S may not even believe or be justified in believing that he/she has that right. But once S *claims* the property, S would have to defend his/her right to it against challenges. Thus, if S *claims* P, S may be called upon to demonstrate his/her right to P. But S may have the right, based on the valid transfer, without being able to produce a defense of it. It is S's claiming the property which imposes upon S the obligation of defending the right to it. Analogously, it is claiming ⌜p⌝ which imposes the obligation to defend the claim. S has the epistemic right to believe that p if the evidence S has is of the appropriate sort; but S may neither believe nor be justified in believing that the evidence is of that sort.

This is simply another way of saying what may seem obvious, namely, that when S asserts ⌐p⌐, S is committed, other things being equal, to asserting that he/she knows that p. In any event, I am willing to grant that much. That assertion, in turn, commits S to asserting that S knows that S knows that p, etc. But, once again, these requirements result from asserting ⌐p⌐, not from knowing that p or believing that p or being justified in believing that p.

Thus Iterative Scepticism is a distinct variety of scepticism which acknowledges that Direct Scepticism may be false. To repeat, Iterative Scepticism asserts that even though S might know that p, S cannot know that S knows that p.

Now, although Iterative Scepticism is a distinct form of scepticism, if Direct Scepticism is correct, Iterative Scepticism is correct. For if Direct Scepticism is true, one of the necessary conditions of knowing that one knows that p cannot be fulfilled—i.e., knowing that p. That is, S could not know that S knows that p, if S could not know that p. But the converse (i.e., Iterative Scepticism entails Direct Scepticism), is not immediately obvious for the reasons already suggested. In fact, I would hold that it is false. For suppose that the argument which is given by the Direct Sceptic for the claim that S can never acquire evidence adequate to confirm the proposition, p, is unsound. Assume further that all the necessary conditions of knowledge can be fulfilled. In that case, of course, Direct Scepticism would be false. But is it not possible that although S can know that p, nevertheless S may fail to know that he/she knows that p? For example, suppose that although S does have adequate confirming evidence for p, S does not have adequate confirming evidence for the claim that he/she has adequate confirming evidence for p. As mentioned before, S may have the epistemic right to believe that p, without having the epistemic right to believe the proposition that *S has the epistemic right to believe that p.* Thus, a person might know that p and yet not be in a position to know that he/she knows that p. I will not argue for that view beyond what has already been said. Rather, I will assume that Iterative Scepticism does not entail Direct Scepticism since it makes the task of refuting Iterative Scepticism more difficult. For without that assumption it would be sufficient to show that Direct Scepticism is false.

Of course, one could reiterate the knowledge predicate indefinitely. And, consequently, there is an infinite variety of scepticisms. But

I will limit the discussion to Direct Scepticism, Iterative Scepticism, and one further variety (discussed below). For the greater the number of iterations the less interesting the sceptic's claim becomes. That interest depends upon the conflict between the initial plausibility of knowledge claims and their denial by the sceptics. It is certainly not initially plausible that on some occasion S knows that S knows that S knows that S knows that p, since it is difficult to think of situations in which such a sentence could have a use. But be that as it may, the arguments used in this essay could be easily converted to refute those less interesting brands of scepticism.

There is a further complication introduced by the Pyrrhonian sceptics. What I will call *Pyrrhonian Direct Scepticism* is the view which refuses to commit itself with regard to the epistemic status of any proposition, including the proposition that S cannot know that p. Pyrrhonian Direct Scepticism asserts that there is no better reason for believing that S can know that p than there is for believing that it is not the case that S can know that p.

It may be thought that there is an even more direct form of Pyrrhonianism — namely, one which claims that there can be no better reasons for believing that p than there are for believing that ∼ p. But if this is a form of scepticism, it would be a form of Direct Scepticism, since the sceptic must be supposing that if there are no better reasons for believing that p than there are for believing its negation, then p is not known. For scepticism is a view with regard to the extent of knowledge, and unless the purported fact of the balance of reasons for and against p provided grounds for some assessment of the extent of our knowledge, it would not be a form of scepticism or provide grounds for a form of scepticism.

There is, at least formally speaking, a Pyrrhonian form of Iterative Scepticism. It would claim that there is no better reason for believing that S can know that S knows that p than there is for claiming that it is not the case that S can know that S knows that p. But the same comments made with regard to the reiterated forms of Iterative Scepticism can be made here. If Pyrrhonian Direct Scepticism can be refuted, Pyrrhonian Iterative Scepticism and other more lengthy forms of iterated Pyrrhonianism seem rather innocuous. Further, the same arguments used here could be used against the forms of iterated Pyrrhonianism.

Thus, the three forms of scepticism that this book will attempt to refute are:

Direct Scepticism	It is not the case that S can know that p.
Iterative Scepticism	It is not the case that S can know that S knows that p.
Pyrrhonian Direct Scepticism	There are no better reasons for believing that S can know that p than there are for believing that it is not the case that S can know that p.

Let me emphasize that I am not claiming that these three forms of scepticism were always distinguished by the sceptics. In fact, as we will see, the seeming plausibility of their accounts sometimes rests on various conflations of these forms of scepticism and the appropriate supporting arguments. However, we will be in a better position to evaluate scepticism if we keep the distinctions between these forms clearly in mind.

1.3 Outline of My Argument

I believe that these three forms of scepticism can be refuted by showing that the following four claims are true:

I   There is no good reason to believe that S never knows that p.

II   There are good reasons to believe that S sometimes knows that p.

III   There is a good argument for the claim that S sometimes knows that p.

IV   There are better reasons for believing that S sometimes knows that p than there are for believing that S never knows that p.

I take it that the conjunction of I and II provides the good argument referred to in III. That is, if I and II are true, then there is a good argument for the claim that S sometimes knows that p. Further, if III is known by S, S has the better reasons referred to in IV. For if I and II do provide S with a good argument for the claim that S knows that p, and if S knows that he/she has a good argument for that claim, then S has better reasons for believing that he/she sometimes knows that p than he/she has for believing that S never knows that p. Thus, Pyrrhonian Direct Scepticism will have been refuted.

Furthermore, if I is correct, there is no reason for accepting Direct

Scepticism, and if II is correct, there are good reasons for rejecting it. And if IV is known by S (on the basis of having read this book, for example), Iterative Scepticism can be rejected. S may not know that he/she knows that p (because, for example, S does not believe that he/she knows that p) but S *can* know that he/she knows that p. For there is a good argument, *knowable* by S, for the claim that S sometimes knows that p. Thus, if S knows that p (and I through III show that there is a good argument for the claim that S sometimes does), and S knows that there is such a good argument, then S has all the evidence needed for the claim that he/she knows that p.

Thus, the core of my argument lies in showing that I and II are correct. That is the task of Chapter Two and Chapter Three respectively. I trust that after those chapters the argument strategy outlined above will become more clear.[7]

1.4 The Context of the Dispute Between the Sceptic and the Nonsceptic

Before taking up the arguments in Chapter Two and Chapter Three, I think that it would be useful to state more carefully what I take to be the context in which the dispute between the sceptic and the nonsceptic arises concerning the extent of our knowledge. For it seems to me that it would be difficult to imagine the dispute proceeding without both parties agreeing to the following points.

The first has already been mentioned. It is that knowledge entails at least true, justified belief. That this is not a set of sufficient conditions for knowledge is painfully obvious—at least, since Gettier's paper.[8] As has already been said, sceptics often argue that knowledge entails, not mere justification, but certainty as well. In fact, the sceptic may argue (and I would think correctly) that the Gettier counterexamples point to cases in which, although a belief is true and justified, it is not certain and, hence, not knowledge. Thus, to ignore this requirement of knowledge is tantamount to ignoring scepticism. I am prepared to grant that knowledge entails absolute certainty.

Second, the sceptic does not (or should not) insist that the propositions expressing the evidence which we have for p entail p. That is, if we let e stand for the evidence, if any, which is adequate for confirming p, then if the sceptic were to require that S knows that p only if e entails p, then the purported dispute vanishes. For it would be readily granted, *without* argument, that the evidence upon which S bases his/her empirical beliefs (for example, that Jones owns a Ford) does not often entail those beliefs.

Nevertheless, it must be granted that occasionally the reasons which we offer for an empirical proposition do entail that proposition. For example, the reasons offered for the proposition that *Jones owns a Ford* may be that *either Jones owns a Ford or Jones stole a Ford* and *Jones did not steal a Ford*. But the sceptic would claim, and and I think rightly so, that unless the propositions which serve as the reasons are themselves justified, they cannot provide a justification of the proposition that *Jones owns a Ford*. But to require that the evidence upon which S bases his/her empirical beliefs must entail those beliefs amounts to requiring that in order for *any* proposition to adequately justify an empirical proposition, it must entail that proposition. In this particular case, it would be to require that the reasons for believing that either *Jones owns a Ford or Jones stole a Ford* and for believing that *Jones did not steal a Ford* must entail those propositions. And the reasons for those reasons must entail them (etc.). I assume that it will be readily granted that many (if not all) of the empirical propositions which we ordinarily take to be knowable are not based upon a chain of reasons *each* link of which is entailed by a preceding one. That is why I say that if the sceptic were to insist that the evidence for an empirical proposition must entail the proposition, the debate would be over before it began. The purported dispute would vanish.

A confusion between fully explaining something and knowing about it may have led some to believe that a necessary condition of knowing that p on the basis of evidence, e, is that e entails p. It *may* be plausible to claim, and some philosophers have claimed, that in some cases in order to fully explain a state of affairs it is necessary that a proposition representing the state of affairs be derivable from another set of propositions representing particular states of affairs and some general lawlike propositions. But that is not a requirement for knowledge. For example, S can know that the phases of the moon have been correlated with the height of the tides without being able to explain that correlation. One can know that there has been such a correlation without being able to derive the proposition representing the correlation from some more fundamental (lawlike) propositions about the effects of the relative positions of the sun, moon, and earth upon the moon's appearance and the earth's oceans. Knowledge that p is one thing; fully explaining p is another.

It could be claimed, however, that this case illustrates situations in which the evidence which we have for a proposition at least nomically implies that proposition.[9] Suppose that S knows that there will be a new moon on Thursday. S could use that information to predict that the tide on Thursday will be the spring (or high) tide of the month. The sceptic could point out that the evidence which S has for the proposition that the tide will be the spring tide for the month is *nomically* implied by the proposition that the moon will be new on Thursday. That is, it could be claimed that there is a nomically necessary connection between the phases of the moon and the height of the tides. For, given the appropriate physical laws, it could not be otherwise than that whenever the moon is new, there is a spring tide. We should grant, I believe, that occasionally our evidence nomically entails the propositions which we believe are knowable.

But, once again, if the sceptic were to require that e either entails or nomically entails p, the argument between the sceptic and non-sceptic vanishes. For it must readily be granted that not every link in the evidential chain terminating with the proposition that the moon will be new on Thursday is an entailment of any kind. If scepticism claimed that we do not know that p simply because some of the links in the evidential chain leading to p are not entailments, then most (if not all) of our purported empirical knowledge is automatically relinquished.

Another fact that may be misleading is that some of our beliefs are justified on the basis of a chain of reasons each link of which is entailed by the preceding one. For if we have a priori knowledge, and if some of the propositions known a priori are deductively inferred from other a priori propositions, then on some occasions the propositions serving as evidence entail the known propositions. But this is not a requirement of empirical knowledge; and it is the possibility of empirical knowledge which is the concern of this book.

Since this is a point that is central to my proposal for removing the plausibility of scepticism, it bears repeating: *If the sceptic insists that knowledge that p on the basis of e is possible only when e entails p, no argument concerning the existence of empirical knowledge can be developed against him/her.* For if there is empirical knowledge that p, it does not rest upon entailing evidence. The challenge of scepticism has been that the arguments which have been developed to support it appear to concern our ordinary empirical beliefs, i.e., beliefs based upon evidence which need not be entailing. As Chisholm says:

> Any adequate theory of evidence must provide for the fact that a proposition *e* may make evident a proposition *h* for a subject S even though e does not entail *h*. We reject the sceptical view according to which there is no reason to believe that the premises of an inductive argument ever confer evidence upon the conclusion. If this sceptical view were true, then we would know next to nothing about the world around us. We would not know, for example, such propositions as are expressed by "There are 9 planets," "Jones owns a Ford," and "The sun will rise tomorrow."[10]

Thus, I take it that there are two general areas of agreement required in order for a dispute to arise between the sceptic and the nonsceptic concerning the extent of our empirical knowledge:

(1) *'S knows that p' entails 'S has a true, justified and certain belief that p'; and*

(2) *In order for S to know that p on the basis of e, it is not required that e entail p.*

The debate between sceptics and nonsceptics can be taken seriously only when the arguments in support of both sides are framed within these areas of agreement.

The purpose of Chapter Two is to show that there are no good reasons which can be adduced within the bounds of these agreements for the claim that S does not know that p. I hope to demonstrate that claim by developing a partial characterization of justification which ought to be acceptable to the sceptic and his/her critics in which to test the various epistemic principles used by the sceptic in arguments for Direct Scepticism. Roughly put, the result of that investigation is that the sceptic's argument either does not provide a reason for accepting Direct Scepticism or it violates the second area of agreement.

The purpose of Chapter Three is to show that there is a good reason for accepting the claim that S can know that p within the two prescribed conditions. Specifically, Chapter Three is designed to defend a characterization of absolute certainty which ought to be acceptable, in principle, to the sceptics, and which is such that many of our empirical beliefs are, in fact, absolutely certain.

CHAPTER 2

THE EVIL DEMON EXORCISED

2.1 The Evil Genius Argument

The task of this chapter is to examine arguments for Direct Scepticism: that is, arguments designed to show that no S can have knowledge that p, where 'p' stands for some empirical proposition which we normally believe is knowable. The number of these arguments is legion; and we cannot examine all of them. But one constantly re-emerges as the last-ditch stand of the Direct Sceptic—it is the Evil Genius Argument. It has been recast for modern readers in the guise of a group of godawful googols, mad scientists equipped with super-computers and vats which contain brains, or other malevolent mechanisms. But the basic structure of the argument remains the same; and if there is no refuge for Direct Scepticism here, I believe that its plausibility will have been removed.

Admittedly, my strategy is to employ a version of the paradigm case argument. For the claim is that the Evil Genius Argument lies at the core of the sceptic's challenge to the possibility of knowledge and that there is no reason to accept Direct Scepticism if the Evil Genius Argument fails. The dangers of such a strategy are mitigated to some extent by the following: it may be incorrect that the Evil Genius Argument is central to every argument for Direct Scepticism; nevertheless, the task of this chapter may yet be accomplished by the strategy employed. This is so because, in the analysis of the Evil Genius Argument, I will focus my comments on the general epistemic principles of justification relied upon by the sceptic in developing the case for Direct Scepticism. If my analyses of these principles are correct, then, if (as I suspect) some other defense of scepticism relied upon them as well, that defense will also have been undermined. Thus, I can ask the reader to examine his/her favorite candidate for a plausible defense of Direct Scepticism in order to determine whether it relies upon epistemic principles similar to those which I hope to show do not provide a basis for Direct Scepticism. As an illustration of this point, I will consider an example of a seemingly different and plausible defense of Direct Scepticism (in section 2.16) which in fact depends upon the same principles which motivate the Evil Genius Argument.

Let me begin our investigation of the Evil Genius Argument by quoting Descartes' famous formulation:[1]

> Nevertheless I have long had fixed in my mind the belief that an all-powerful God existed by whom I have been created such as I am. But how do I know that He has not brought it to pass that there is no earth, no heaven, no extended body, no magnitude, no place, and that nevertheless [I possess the perceptions of all these things and that] they seem to me to exist just exactly as I now see them? And, besides, as I sometimes imagine that others deceive themselves in the things which they think they know best, how do I know that I am not deceived every time that I add two and three, or count the sides of a square, or judge of things yet simpler, if anything simpler can be imagined? . . .

> I shall then suppose, not that God who is supremely good and the fountain of truth, but some evil genius not less powerful than deceitful, has employed his whole energies in deceiving me; I shall consider that the heavens, the earth, colours, figures, sound and all other external things are nought but the illusions and dreams of which this genius has availed himself in order to lay traps for my credulity; I shall consider myself as having no hands, no eyes, no flesh, no blood, nor any senses, yet falsely believing myself to possess all these things, I shall remain obstinately attached to this idea, and if by this means it is not in my power to arrive at the knowledge of any truth, I may at least do what is in my power (i.e. suspend my judgment), and with firm purpose avoid giving credence to any false thing, or being imposed upon by this arch deceiver, however powerful and deceptive he may be.

It is not clear from this passage whether Descartes is defending Direct Scepticism, Iterative Scepticism, or even Pyrrhonian Direct

Scepticism. Is he claiming that because (at least, at a certain stage in the *Meditations*) he cannot eliminate the possibility that the evil genius is deceiving him, he cannot know what he formerly believed that he had known (a form of Direct Scepticism)? Is he claiming that because he cannot eliminate the possibility of being deceived, he cannot know whether he knows (a form of Iterative Scepticism)? Or finally, is he claiming that because he cannot eliminate the possibility of being deceived, he has no better reason for believing that he knows than that he is mistaken and hence he should "suspend judgment" (a form of Pyrrhonian Direct Scepticism)? Or is he, perhaps, claiming that all three forms of scepticism follow from the fact that he cannot eliminate the possibility of an evil demon making him falsely believe those things which he thought that he knew?

Of course, the *most* important question to ask is: why should the fact (if indeed it is a fact) that Descartes cannot eliminate the possibility of an evil genius lead to any form of scepticism? What are the general epistemic principles to which he is appealing? For example, is he implicitly claiming that in order for a person, S, to know that p on the basis of some evidence, e, e must provide S with evidence which is adequate for eliminating all of the other contrary propositions to p which are consistent with e? Or is he appealing to a weaker epistemic principle, namely, that if S is justified in believing that p, then S is justified in denying any contrary of p?

The passage from Descartes does not provide clear answers to these questions. But, as I mentioned, the argument has its modern defenders, and perhaps some of these questions can be answered after examining a contemporary version. Both Keith Lehrer[2] and Peter Unger[3] avail themselves of it. I will quote Lehrer's presentation of the argument at some length and return to it frequently in this chapter with the hope of clarifying just what it is that is claimed by the sceptic. There will be occasions to refer to Unger's version as well.

> Now it is not at all difficult to conceive of some hypothesis that would yield the conclusion that beliefs of the kind in question are not justified, indeed, which if true would justify us in concluding that the beliefs in question were more often false than true. The sceptical hypothesis might run as follows. There are a group of creatures in another galaxy, call them Googols, whose intellectual capacity is 10^{100} that of man, and who amuse themselves by sending out a peculiar kind of wave that affects our brain in

such a way that our beliefs about the world are mostly incorrect. This form of error infects beliefs of every kind, but most of our beliefs, though erroneous, are nevertheless very nearly correct. This allows us to survive and manipulate our environment. However, whether any belief of any man is correct or even nearly correct depends entirely on the whimsy of some Googol rather than on the capacities and faculties of the man. . . . I shall refer to this hypothesis as the *sceptical hypothesis*. On such a hypothesis our beliefs about our conscious states, what we perceive by our senses, or recall from memory, are more often erroneous than correct. Such a sceptical hypothesis as this would, the sceptic argues, entail that the beliefs in question are not completely justified. . . . In philosophy a different principle of agnoiology is appropriate, to wit, that no hypothesis should be rejected as unjustified without argument against it. Consequently, if the sceptic puts forth a hypothesis inconsistent with the hypothesis of common sense, then there is no burden of proof on either side, but neither may one side to the dispute be judged unjustified in believing his hypothesis unless an argument is produced to show that this is so. If contradictory hypotheses are put forth without reason being given to show that one side is correct and the other in error, then neither party may be fairly stigmatized as unjustified. However, if a belief is completely justified, then those with which it conflicts are unjustified. Therefore, if neither of the conflicting hypotheses is shown to be unjustified, then we must refrain from concluding that belief in one of the hypotheses is completely justified. . . .

Thus, before scepticism may be rejected as unjustified, some argument must be given to show that the infamous hypotheses employed by sceptics are incorrect and the beliefs of common sense have the truth on their side. If this is not done, then the beliefs of common sense are not completely justified, because conflicting sceptical hypotheses have not been shown to be unjustified. From this premise it follows in a single step that we do not know those beliefs to be true because they are not completely justified. And then the sceptic wins the day.[4]

I think that it is clear that Lehrer is defending a version of Direct Scepticism, since he claims that our inability to reject the sceptical hypothesis leads to the conclusion that our beliefs are not justified and, thereby, not known. I assume that he would also endorse Iterative Scepticism because, as mentioned in section 1.2, Direct Scepticism entails Iterative Scepticism. In addition, I would suspect that he would reject Pyrrhonian Direct Scepticism, since he thinks that our inability to reject the sceptical hypothesis provides us with a better reason for believing that Direct Scepticism is correct than that it is incorrect.

But even this modern version of the Evil Genius Argument has two fundamentally different interpretations. For the sceptical hypothesis may be interpreted in either of two rather different ways. It could be interpreted to mean that the evil genius, googol, or other malevolent mechanism is, *in fact,* bringing it about that our belief that p is false (i.e., making it *only appear* that p) or it might mean that there is such a malevolent mechanism which *could* bring it about that our beliefs are false (and it remains unsettled and unimportant whether it is *in fact* exercising its power). Now, of course, the latter interpretation is entailed by the former, but the converse is not true. And as we will see, there are *crucial differences* between the sceptical epistemic principles motivating the two interpretations of the hypothesis.

Before considering these two interpretations in the next section, let me point to two important features of the argument. The first is that the sceptical argument appears to be framed within the areas of agreement mentioned earlier, namely, that knowledge entails true, justified (and certain) belief and that the evidence for p need not entail p. For the sceptic is not arguing that because e fails to entail p, p is not known. It is assumed that e does not entail p, but it is not *that* fact which leads to Direct Scepticism; it is rather the purported fact that S is not justified in rejecting the sceptical hypothesis which leads to scepticism. In addition, the sceptic is willing to grant that e is true. That is, the sceptic seems willing to grant that we have memories, but that they are "more often erroneous than correct"; he/she seems willing to grant that we have sensory perceptions, but that they are "more often erroneous than correct"; etc. Now if e were to entail p, then the sceptical hypothesis (on either interpretation) would be false since it would not be *possible* for p to be false when e is true. (Let me note parenthetically that the sceptic is, of course, not granting that e is known.)

The second point is that the epistemic principles embodied in the Evil Genius Argument can also be found in arguments for scepticism which do not explicitly invoke an epistemically malevolent mechanism. For example, consider this argument cited by Cicero concerning what he calls the "famous Servilius twin of the old days":[5]

If therefore a person looking at Publius Servilius Geminus used to think he saw Quintus, he was encountering a presentation of a sort that could not be perceived, because there was no mark to distinguish a true presentation from a false one; and if that mode of distinguishing were removed, what mark would he have, of such a sort that it could not be false, to help him to recognize Gaius Cotta, who was twice consul with Geminus? You say that so great a degree of resemblance does not exist in the world. You show fight, no doubt, but you have an easy-going opponent; let us grant by all means that it does not exist, but undoubtedly it can appear to exist, and therefore it will cheat the sense, and if a single case of resemblance has done that, it will have made everything doubtful; for when that proper canon of recognition has been removed, even if the man himself whom you see is the man he appears to you to be, nevertheless you will not make that judgement, as you say it ought to be made, by means of a mark of such a sort that a false likeness could not have the same character. Therefore seeing that it is possible for Publius Geminus Quintus to appear to you, what reason have you for being satisfied that a person who is not Cotta cannot appear to you to be Cotta, inasmuch as something that is not real appears to be real?[6]

Roughly, the argument contained in this passage appears to be that if a presentation is to be of the type which yields knowledge, it must be such that no instance of the type can be false. But since twins resemble each other so closely or at least can appear to resemble each other so closely, the presentation of one of them may be indistinguishable from that of the other. Hence, the presentation of Publius and that of Quintus have no internal "mark" by which they can be distinguished. Even in cases in which it is known that no twin exists (e.g., Cotta in the paragraph quoted from Cicero), there is no internal mark of the presentation which allows us to say that the presentation is such that it could not be false. As far as the presentations go, *everything* of which we have a presentation *could* have a twin, and, thus, the presentation could be false.

The sceptic is not denying that our presentations have internal marks which serve as "evidence" for our beliefs. Nevertheless, the sceptic is arguing that the evidence is consistent with the hypothesis that twins are or may be ubiquitous. The characteristics of every presentation are such that it is possible that every one of them could be false. In order to be justified in believing that we are seeing Quintus (my watch, a book, etc.), we must be justified in denying that it is not Publius (a watch-facsimile, a book-facsimile, etc.). Thus, at the core of this argument is the sceptical hypothesis, although there is no appeal to a personified, all-powerful, malevolent being which made the sceptical hypothesis immediately understandable to seventeenth-century Europeans. Nor is there a reference to a race of superscientists armed with mighty computers and fancy vats in which to sustain our brains—the science fiction version of the hypothesis which seems to render it immediately understandable to us. But the argument is essentially the same; its general form can be put as follows:

> There is at least one understandable hypothesis which we must be justified in denying in order for us to be justified in believing those propositions which we believe are knowable. We cannot be justified in denying that understandable hypothesis. Thus, we cannot be justified in believing those propositions which we normally believe are knowable. And since justification is a necessary condition of knowledge, we cannot have the knowledge which we normally believe that we do have.

The appeal to the "famous Servilius twins" of Cicero, or the godlike evil genius of Descartes, or the science fiction creatures of contemporary sceptics is designed to produce an immediately understandable sceptical hypothesis. The particular mechanism mentioned by the sceptic in the sceptical hypothesis is chosen because of its heuristic value and is unimportant for assessing the soundness of this basic argument. I will continue to call this fundamental argument The Evil Genius Argument and the sceptical hypothesis contained in it the Evil Genius Hypothesis. But since my strategy will be to investigate the variety of general epistemic principles embodied in the argument, the conclusions reached here should apply to all of those sceptical arguments which share this common form but do not appeal explicitly to the evil genius.

2.2 The Sceptical Hypotheses and Basic Epistemic Maxim Which Result in Four Sceptical Epistemic Principles

As noted in the previous section, the sceptical hypotheses occur in two related but, nevertheless, distinct forms. In the first hypothesis, which I will call the *Contraries Hypothesis* (H_c), some malevolent mechanism is postulated which actually brings it about that our purported knower, S, falsely believes that p, even though all of S's evidence, e, for p remains the same. In the second hypothesis, the *Defeaters Hypothesis* (H_d), a mechanism is postulated which *could* bring it about that S's beliefs are false even though all of S's evidence, e, for p remains unchanged.

The two versions of the sceptical hypothesis can be put as follows:

Contraries Hypothesis (H_c)	e & ~p & there is some mechanism, M, which brings it about that S believes (falsely) that p.
Defeaters Hypothesis (H_d)	e & there is some mechanism, M, which could bring it about that S falsely believes that p without changing the truth of 'e'.[7]

Of course, the third conjunct of H_c entails the second, thus making the latter redundant, but I list '~p' separately in order to emphasize the important distinction between H_c and H_d. That difference is that while H_c and p are contraries (both can be false, but both cannot be true), H_d and p are not contraries (they can both be true). Both H_c and H_d postulate the existence of M, the malevolent mechanism. It is its sheer power in H_d and the exercise of it in H_c which distinguishes H_c from H_d.

An initial formulation of what I will call the sceptic's *Basic Epistemic Maxim* which occurs in the general form of the Evil Genius Argument could be put as follows: *S must be justified in rejecting H (H_c or H_d) in order to be justified in believing that p.* As we shall see, this maxim is ambiguous in a variety of ways, and our task will be to evaluate each version of it in order to determine whether any version provides a basis for Direct Scepticism. In fact, a major task of this chapter will be to disambiguate the Basic Epistemic Maxim. Roughly put, we will find that some interpretations of this basic maxim yield valid epistemic principles which are useless to the sceptic; and other interpretations yield false epistemic principles which would be useful if true. It is the conflation of these various interpretations which provides scepticism with its initial plausibility.

The *first* interpretation of this Basic Epistemic Maxim takes the "in order to be" as designating a logical consequence; that is: S is justified in believing that p *logically implies* that S is justified in rejecting H (H_c or H_d). Given this interpretation, the sceptic is asserting a *logical* relationship between S's being justified in believing that p and S's being justified in denying that H (H_c or H_d). The sceptic continues, often implicitly, by claiming that S is not justified in rejecting H (H_c or H_d) and, since S is not justified in believing that p, S does not know that p. Viewed this way, Descartes' answer to his own formulation of the sceptical argument was to demonstrate that

S is justified in rejecting H since there is a "proof" of God's existence and "epistemic" benevolence.

On this first interpretation of the Basic Epistemic Maxim, the sceptic appears to be relying upon either of two different and general epistemic principles of justification depending upon whether H is construed as H_c or construed as H_d. The sceptic could be asserting that if x and y are contraries and if S is justified in believing that x, then S is justified in denying that y; or the sceptic could be asserting that if y is a defeater of the justification of x and if S is justified in believing that x then S is justified in denying that y. The notion of a defeater will be explained in detail in Chapter Three, but roughly I mean that y is a defeater of the justification of x, if and only if, when y is conjoined with whatever justified x for S, the resulting conjunction no longer justifies x for S. If we are to determine whether the sceptic has provided a *reason* for accepting the claim that S cannot know that p on the basis of this interpretation of the Basic Epistemic Maxim, we must examine the two principles inherent in this first interpretation in order to ascertain whether either of them provides an acceptable basis for Direct Scepticism.

The *second* interpretation of the sceptic's Basic Epistemic Maxim takes the "in order to be" not merely as designating a logical *consequence* of justification, but rather as delineating an evidential *prerequisite* of justification. Hence, whereas the first interpretation would assert that S is justified in denying that y whenever S is justified in believing that x (where x and y are contraries), the second interpretation of this basic maxim would require that S be justified in denying that y as a necessary step toward being justified in believing that x. In other words, S's justification of x consists, at least in part, in the elimination of the contraries of x.

Since there are two versions of the sceptical hypothesis, H_c and H_d, two general epistemic principles result from this second interpretation of the Basic Epistemic Maxim. In addition to the one just presented, the sceptic could be claiming that S must first eliminate all of the defeaters of a justification of a proposition before that proposition is justified.

Thus the sceptic's Basic Epistemic Maxim can be interpreted to mean any of the following four epistemic principles:

Contrary Consequence Elimination Principle	For any propositions, x and y, (necessarily) if y is a contrary of x, then if S is justified in be-

	lieving that x, then S is justified in believing that not y.
Defeater Consequence Elimination Principle	For any propositions, x and y, (necessarily) if y is a defeater of the justification of x, then if S is justified in believing that x, then S is justified in believing that not y.
Contrary Prerequisite Elimination Principle	For any propositions, x and y, (necessarily) if y is a contrary of x, then if e is adequate evidence for x, then e contains not y.
Defeater Prerequisite Elimination Principle	For any propositions, x and y, (necessarily) if y is a defeater of the justification for x, then if e is adequate evidence for x, e contains not y.

The remainder of Chapter Two is a consideration of these four epistemic principles.

2.3 Some Abbreviations Used Throughout this Book

Before proceeding to examine the epistemic principles used by the sceptic, it will be useful to introduce some shorthand. I will use the following abbreviations throughout the rest of this book.

S	a person.
p	any one of a number of contingent propositions believed to be knowable, e.g., I have a hand, Jones owns a Ford, no person has yet eaten an adult elephant at one sitting, the phases of the moon are correlated with the heights of the tides, some people enjoy Bach's music more than Wagner's music.
e	the evidence for p.
xCy	x confirms y; i.e., according to the rules of confirmation, it is permissible to infer y from x.
x$\mathcal{C}$y	x fails to confirm y; i.e., according to the rules of confirmation, it is not permissible to infer y from x.
Jsx	S has a justification for x or S is entitled to believe that x.
Js~x	S has a justification for ~x or S is entitled to deny x.
Gsx	x is grounded for S.
Osux	u is an overrider of the confirmed proposition, x, for S.
H_c	e & ~ p & there is some mechanism, M, which brings it about that S believes (falsely) that p.
H_d	e & there is some mechanism, M, which could bring it about that S falsely believes that p (without changing the truth of 'e').
H	$H_c \vee H_d$

2.4 Two of the Sceptical Epistemic Principles: The Contrary Consequence Elimination Principle and the Defeater Consequence Elimination Principles

We will begin our examination of the principles used by the sceptic in the Evil Genius Argument by considering the two *consequence principles.* Those two principles interpret the Basic Epistemic Maxim to mean that there is a logical relationship between being justified in believing that p and being justified in denying that H. Recall that there are two distinct versions of H and that H_c and p are contraries, but that H_d and p are not contraries. The two consequence principles are:

Contrary Consequence Elimination Principle	For any propositions, x and y, (necessarily) if x and y are contraries, then $Jsx \rightarrow Js \sim y$.
Defeater Consequence Elimination Principle	For any propositions, x and y, (necessarily) if y is a defeater of the justification of x, then $Jsx \rightarrow Js \sim y$.

We must now ask whether either of these consequence principles is acceptable. If not, is some revision of them acceptable? And, further, if some form of them is acceptable, does that form provide grounds for Direct Scepticism? Briefly, my answer is that only the Contrary Consequence Elimination Principle is acceptable, but that neither it nor the Defeater Consequence Elimination Principles can provide the grounds necessary for Direct Scepticism. But that is the end of the journey, and, although this journey is not a thousand miles long, it does proceed step by step; so let us begin by examining the Contrary Consequence Elimination Principle.

2.5 A Defense of the Contrary Consequence Elimination Principle

The Contrary Consequence Elimination Principle (or some modified version of it) certainly *seems* true. One might even consider it a standard of rationality. For it would seem that if it is not possible for two propositions to be true, then if S is justified in believing that one of them is true, S has available a justification for rejecting the other. We might want to add the requirement that S have the concepts necessary to understand both propositions (à la Chisholm[8]), or, perhaps, some other minor modification might be called for. It does, at least at first glance, *seem* acceptable.

But in spite of its intuitive appeal, some philosophers have argued that the Contrary Consequence Elimination Principle cannot provide any grounds for Direct Scepticism. They do so, not because of the considerations developed later in this chapter, but rather because they believe that this principle (even with the appropriate restrictions to be considered later) is false. Of course, if this principle were false, it would make the task of refuting Direct Scepticism much easier, but I believe that the arguments against it are inconclusive. In fact, I believe that there is a convincing argument for a stronger epistemic principle, namely, if x entails y and if S is justified in believing that x, then S is justified in believing that y. I will call this stronger principle the *Principle of the Transmissibility of Justification Through Entailment* following Irving Thalberg.[9]

Since considerable time will be devoted to defending the stronger principle and its corollary, the Contrary Consequence Elimination Principle, it may be wondered why we cannot simply grant that step in the argument to the sceptic. Some defense of my procedure seems necessary.

I believe that the examination of this step in the sceptic's argument is not only useful but necessary for the following reasons. First, by showing that the arguments against this interpretation of the sceptic's Basic Epistemic Maxim are inconclusive, scepticism becomes more plausible, and, hence, the refutation of it all the more important. Second, my claim in Chapter One that scepticism relies upon the traditional analysis of knowledge as true, justified belief will become more clear and acquire some additional support, since my defense of the Contrary Consequence Elimination Principle will rest in large part upon a discussion of epistemic justification. Third, our examination of the arguments against the Contrary Consequence Elimination Principles will prove useful when we turn to an examinatin of the Defeater Consequence Elimination Principle and the other interpretations of the sceptic's Basic Epistemic Maxim. To look ahead a bit, we will find that although the arguments against the Contrary Consequence Elimination Principle miss the mark, they can be redirected against the Defeater Consequence Elimination Principle. Fourth, the partial analysis of justification developed in defense of the stronger Principle of Transmissibility will provide a basis for the discussion of knowledge and certainty in Chapter Three. Fifth, since the Gettier-related problems with the traditional analysis of knowledge depend upon a version of the stronger principle, by showing that it is correct, I will have presented an indirect argument against those who believe that the Gettier issues arise from unacceptable epistemic principles.[10] Finally, by supporting the Contrary Consequence Elimination Principle, I will be following the policy adopted earlier—i.e., granting to the sceptic all that is reasonably possible to grant.

Thus, the examination of the arguments against the Contrary Consequence Elimination Principles is not a digression from our main task. It is a necessary step in my proposed refutation of scepticism. The sceptical position will be understood more clearly if we investigate the relevant features of the concepts of justification relied upon by the sceptic. Furthermore, some of the results reached here will prove useful when we discuss the relationship between certainty and knowledge in Chapter Three.

My consideration of the Contrary Consequence Elimination Principle has two steps. First, I wish to show that the objections to it are not well founded. Second, I wish to present an argument for the stronger Principle of Transmissibility based upon a proposed model of justification which I hope will be acceptable to both sceptics and nonsceptics.

Let us begin by considering the arguments against the Contrary Consequence Elimination Principle. Fred Dretske presents what I take to be the relevant, critical objection to that principle.[11] Others, most notably Irving Thalberg, have also argued against the transmissibility of justification through entailment;[12] but since Dretske's argument was framed with scepticism in mind, I will concentrate on his formulation of the objections to it.

The argument against the Contrary Consequence Elimination Principle is based upon several examples of situations in which Dretske claims that even though some proposition, say p, entails another, say q, and S is justified in believing that p, S is not justified in belieiving that q. Dretske's argument is directed against the stronger transmissibility principle, but as the example about to be quoted shows, he also believes that the Contrary Consequence Elimination Principle is invalid. For, in the example cited, the entailed proposition is the denial of a contrary. All the examples are variations on one theme, and, if I am correct, the mistake which Dretske makes in analyzing them is common to all. Let us call this case the *Zebra Case*:

You take your son to the zoo, see several zebras, and, when questioned by your son, tell him they are zebras. Do you know they are zebras? Well, most of us would have little hesitation in saying that we did know this. We know what zebras look like, and, besides, this is the city zoo and the animals are in a pen clearly marked "Zebras." Yet, something's being a zebra implies that it is not a mule and, in particular, not a mule cleverly disguised by the zoo authorities to look like a zebra. Do you know that these animals are not mules cleverly disguised by the zoo authorities to look like zebras? If you are tempted to say "Yes" to this question, think a moment about what reasons you have, what evidence you can produce in favor of this claim. The evidence you *had* for thinking them zebras has been effectively neutralized, since it does not count toward their *not* being mules cleverly disguised to look like zebras. Have you checked with the zoo authorities? Did you examine the animals closely enough to detect such a fraud? You might do this, of course, but in most cases you do nothing of the kind. You have some general uniformities on which you rely, regularities to which you give expression by such remarks as, "That isn't very likely" or "Why should the zoo authorities do that?" Granted, the hypothesis (if we may call it that) is not very plausible, given what we know about people and zoos. But the question here is not whether this alternative is plausible, not whether it is more or less plausible than that there are real zebras in the pen, but whether you know that this alternative hypothesis is false. I don't think you do.[13]

Thus Dretske claims to have discovered a set of cases (remarkably similar to the standard Evil Genius Case) in which S is justified in believing that p ('there are zebras in the zoo') and p → q (where q is 'the zebra-looking animals are not cleverly diguised mules'), but in which S is not justified in believing that q. As I said before, if he is correct that the Contrary Consequence Elimination Principle is invalid, the sceptic will have lost an important weapon in his/her arsenal used to defend Direct Scepticism.

Dretske's strategy, then, for refuting this principle is to search the reasons which S has which justify the belief that p — call those reasons w_1 — and to show that those reasons do not justify S in believing that q. Thalberg sees the issue similarly. In fact, one of his formulations of the principle under discussion is:

For any proposition P, if a person S is justified [by evidence propositions E_1 . . . E_n which S accepts] in believing P, and P entails Q, and S deduces Q from P and accepts Q as a result of this deduction, then S is justified [by E_1 . . . E_n] in believing Q.[14]

Thalberg has added the critical material contained in the brackets. The rest, as he says, comes from Gettier's famous essay. My point here is merely that Thalberg construes the issue in a manner similar to that implicit in the purported counterexample to the Contrary Consequence Elimination Principle developed by Dretske. Specifically, the general question concerning the transmissibility of justification through entailment has been interpreted more narrowly by imposing the restriction that if p entails q and e justifies p for S, then e must justify q for S. With this restriction, the strategy becomes simply to show that the evidence ('E_1 . . . E_n' in the quote from Thalberg's article) which is adequate for p is not adequate for q.

Now, I am willing to grant that the reasons, w_1, which S has for believing that p do not provide a justification for q; that is, I will grant that $w_1 \mathcal{C} q$. But that is not sufficient to establish the desired conclusion. For does it follow from the fact that $w_1 \mathcal{C} q$ that S is not justified in believing that q? Only if there are no *other* reasons available to S which would provide adequate evidence for q. That is, in order to show that S is not justified in believing that q, it must be shown that none of the other reasons which S *must* have are adequate to justify the belief that q.

Dretske takes partial cognizance of this when he adds the stipulation that S has not taken "special precautions" to guard against error in this case. That is, S has no reasons other than the ones specifically attributed to him/her in explication of the Zebra Case. Now, I am willing to grant that as well. Thus, I am willing to grant that neither w_1 nor any "special" evidence is available to S which would provide adequate grounds for q. But, once again, we must ask: does it follow that S is not justified in believing that q? Only if there is no other reason which *must* be available to S which provides adequate confirming evidence for q.

I submit that there surely is such a proposition. In fact, I submit that there is a *conclusive* reason available to S. That reason, of course, is p, itself! For if S has a justification for p, then it would seem that p, itself, is available for S to use as a reason for further beliefs. In addition, since p → q, p provides S with an *extremely good* justification for q. In fact, what better reason than p could S wish to have for q?

Now, if I am correct here, it may be wondered why neither Dretske nor Thalberg considered p as a possible candidate for an adequate reason for q. I suspect that the explanation depends upon the way in which they have construed both the Principle of the Transmissibility of Justification and the Contrary Consequence Elimination Principle. Let us concentrate on the stronger of the two. The transmissibility principle could be put as follows:

Principle of the Transmissibility of Justification Through Entailment

For any propositions, x and y, (necessarily) if x → y and Jsx, then Jsy.

It seems to me that Dretske and Thalberg have implicitly assumed that the truth of the transmissibility principle depends upon the truth of another principle, which we will call the *Partial Transitivity of Confirmation Principle:*

Partial Transitivity of Confirmation Principle

For any propositions, x and y and z, (necessarily) if x → y and zCx, then zCy.

They have argued that this Partial Transitivity of Confirmation Principle is false by finding a particular case in which some evidence, w_1, does not confirm a proposition, say q, but w_1 does confirm p and p does entail q. More specifically, the evidence that the animals in the zoo in a pen marked "Zebras" look like zebras does not confirm the proposition that they are not cleverly disguised mules, but that evidence does confirm the claim that they are zebras and the proposition that *they are zebras* does entail the proposition that *they are not cleverly disguised mules.* Since they believe that the truth of the transmissibility principle depends upon the truth of the Partial Transitivity of Confirmation Principle, and the latter is false, there appears to be no need to consider whether there is some proposition other than w_1 which confirms q. Thus, whether p confirms q is irrelevant.

Now, my argument up to this point has been merely that the truth of the transmissbiility principle (and, hence, of its corollary, the Contrary Consequence Elimination Principle) does not depend upon the truth of the Partial Transitivity of Confirmation Principle. Put another way, the truth of the transmissiblity principle does not depend upon whether the evidence which confirms the antecedent in an entailment for S also confirms the consequent in the entailment for S, but, rather, upon whether whenever there is evidence which confirms the antecedent for S, S has *some* evidence which confirms the consequent. The Principle of Transmissibility does not require that *the* evidence which confirms the consequent must be *the very same* evidence which confirms the antecedent.

In reply, it could be granted that I am correct in believing that whether p confirms q is relevant to an assessment of the transmissibility principle, but it may be objected that there are good reasons which prevent me from claiming that p does confirm q for S. I would like to consider three such alleged good reasons. The first depends upon a mistaken view concerning the transitivity of confirmation; the second fails to recognize the implication of the claim that the reasons for p have been "neutralized"; and the third conflates the distinction between the Contrary Consequence and the Defeater Consequence Elimination Principles.

Let us examine the first objection in some detail before turning to the second and third. It should be remembered that my primary purpose in this and the three following subsections is to support the Contrary Consequence Elimination Principle in order to demonstrate the power and attractiveness of the arguments for Direct Scepticism. But these issues are interesting in their own right, and the discussion here provides a basis for the constructive task of developing an acceptable account of certainty—the task of Chapter Three. So let us turn to the first objection to using p as the confirming evidence which S has for q.

The objection could be put this way: since I have granted that $w_1 Cp$ (as stipulated in the Zebra Case), if I were to insist that pCq, then I must also grant that $w_1 Cq$. But I have already granted that $w_1 \mathcal{C}q$ (w_1 fails to confirm q) and, thus, I cannot consistently claim that pCq.

This would be a conclusive objection to counting p as confirming evidence for q only if the following principle were true:

Principle of the Transitivity   For all x and y and z, (necessarily) if
of Confirmation xCy and yCz, then xCz.

I wish to show that this principle is false. As I just mentioned, it will be useful to do so because it will help to clarify and support the sceptic's position by showing that an adequate defense against the Dretske-like attack depends upon the nontransitivity of confirmation. In other words, if confirmation were transitive, then p could not confirm q (in the Zebra Case), since it is granted that $w_1 Cp$ and $w_1 \mathcal{C}q$. And if p cannot confirm q, then the Zebra Case will serve as a counterexample to the Contrary Consequence Elimination Principle — a principle upon which Direct Scepticism is based. Put another way, if confirmation were transitive, the Zebra Case would show that the Principle of the Transmissibility of Justification Through Entailment is false.

I have argued elsewhere that justification is not a transitive relation,[15] and a similar argument can be used here to show that confirmation is not transitive. In fact, it will be seen later that *since* confirmation is not transitive, neither is justification (see section 2.8). Consider a case in which Jones, a clever car thief, has stolen a car. He knows that the best way to get away with the theft is to behave as much like a car owner as possible. Jones knows that car owners usually drive their cars, garage them, have valid-looking titles, speak about owning their cars, have friends who will testify to their ownership, etc. Jones is in a position to bring all of that about. Thus, the proposition that Jones drives a car, garages it, has a valid-looking title to it, etc., seems confirmed. And the proposition that Jones drives a car, garages one, etc., confirms the proposition that Jones owns a car.

Let us represent this case as follows:

w' Jones is a clever car thief and has stolen a car. Jones knows that it is best to behave as much as possible the way car owners behave. Jones identifies a set of car-owning behaviors which Jones is in a position to duplicate and which will serve as adequate evidence for the claim that Jones owns a car (e.g., driving and garaging the car, possessing a valid-looking title to it).

p' Jones behaves in that fashion: Jones drives and garages the car and has a valid-looking title, etc.

q' Jones owns a car.

Now, clearly it is possible that $w'Cp'$ and $p'Cq'$, but also that $w'\mathcal{C}q'$. For in this case w' contains some evidence which overrides the inference from p' to q'.

As this example illustrates, I am using expressions of the form 'x confirms y' in a somewhat nontypical fashion. It would probably be more "natural" to say that the *facts that James stole a car and pretended to own it confirm the hypothesis* that Jones is a clever car thief. That is, it is facts or data which are typically said to confirm propositions or hypotheses. However, as I use the expression, both the "x" and the "y" in 'x confirms y' are taken to range over propositions.

Within the bounds of ordinary language, one could, perhaps, say that if S is justified in believing that Jones is a clever car thief, then S is justified in believing that Jones pretends to own a car. Consequently, a "natural" extension of ordinary linguistic practice would be to claim that the proposition 'Jones is a clever car thief' justifies the proposition 'Jones pretends to own a car'. In other words, it may seem more appropriate to use 'x justifies y' rather than 'x confirms y'.

But such a choice would tend to obscure a distinction which I think is crucial to maintain. For on the one hand, there is the deductive or inductive relation between sets of propositions—one set rendering the other set evident. While on the other hand, there is the relation between a person and a set of propositions—the set of propositions being evident to a person. We ordinarily use 'justification' to designate both relations when we say that S is *justified* in believing that p on the basis of some evidence, e, because e *justifies* p. In order to avoid conflating these two meanings of 'justification,'I will use the expression 'x confirms y' to designate the relation between propositions x and y such that it is epistemically permissible to infer y from x. This point could be put in a slightly different fashion: As I use the expressions 'justification', 'knowledge' and 'confirmation', confirmation is a constituent of justification which is a constituent of knowledge.

Even with this explanation of the use of 'confirms', some philosophers may be worried about the tenses in the Clever Car Thief Case.[16] That is, they may think that what is rendered evident or confirmed by w' is not p', but rather p^*, where p^* is *Jones will behave in a car-owning fashion at some time,* t_f (where t_f is some time in the future). But, then, since p^* confirms q^*, where q^* is *Jones will own a car at* t_f, the example, even interpreted in this fashion, shows that confirmation is not transitive since $w'Cp^*$ and p^*Cq^*, but $w'\mathcal{C}q^*$.

We will encounter other cases in which confirmation fails to be transitive, but I think that we can already point to one general cause of the failure of transitivity. It depends upon one of the areas of agreement reached between the sceptic and the nonsceptic at the very outset of the discussion, namely, e can confirm p even though e does not entail p. That means that confirmation is an overridable relationship. Thus, even though eCp, there may be another proposition, say o, which is such that $(e\&o)\mathcal{C}p$. In the case just considered, the proposition *Jones drives a car, garages it, and has a valid-looking title to it* confirms the proposition *Jones owns a car.* But the proposition *Jones is a clever car thief, has stolen a car, and as such duplicates car-owning behavior* is an overriding proposition. In this case, the very same proposition, o, acts as both the *confirming* evidence for e *and* the *overriding* evidence of the proposition confirmed by e. Thus, in this case, oCe, eCp, but o𝒞p because o counts, so to speak, for e but also against p.

Of course, not all cases in which transitivity fails are cases in which there is an overrider present. The Zebra Case is an example of one in which w_1 Cp and pCq, but w_1 𝒞q, not because w_1 counts against q but rather because the rules of confirmation simply do not include a permissible inference from w_1 to q. The rules allow an inference from w_1 to p and from p to q, but they do not permit an inference from w_1 to p.

Let me illustrate this last point by considering another typical rule-governed situation. Chess rules are often taken to display some typical features of rule-governed situations. Consider a chessboard with the squares identified as follows:

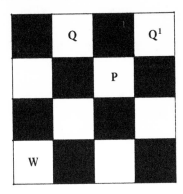

Figure 1.

Suppose that 'xBy' means that according to the rules of chess it is permissible to move a bishop to y from x (where 'x' and 'y' range over chessboard squares). I take this to be analogous to the issue here; for 'xCy' means that it is permissible according to the rules of evidence to infer y from x (where 'x' and 'y' range over propositions). Now, just as it is permissible for the bishop to move from W to P and from P to Q (or Q^1), and it is not permissible for it to move from W to Q (although it would be to move from W to Q^1), it can be permissible to move (inferentially) from w_1 to p and from p to q, but not permissible to move from w_1 to q.

Thus, this first objection to counting p as the adequate evidence for q in the Zebra Case can be discounted, since it depends upon the mistaken belief that adequate confirming evidence is a transitive relationship. Specifically, the evidence, w_1, which S has for the proposition that there is a zebra in the zoo, need not, itself, be adequate confirming evidence for the proposition that q (the animal believed to be a zebra is not a cleverly disguised mule), in order for S to have adequate evidence for q even when p confirms q.

Let us now turn to the second general objection to considering p as the adequate confirming reason for q in the Zebra Case. In spite of what I have said thus far, it may appear that Dretske has presented a good reason for rejecting both the Contrary Consequence Elimination Principle and the stronger Principle of the Transmissibility of Justification Through Entailment. Recall that Dretske claimed that w_1, the evidence for p, has been "neutralized." It could be claimed that if w_1 has been neutralized, anything evidentially dependent upon w_1 (p and q) has also been neutralized. Thus, p could not be used by S to confirm q because p has been "neutralized."

Now, I believe that it would be correct to claim that if w_1 has been "neutralized," q is no longer justified. But this is the case not because p fails to confirm q or because the Contrary Consequence Elimination Principle is false, but rather because *p is no longer justified for S*!

Consider the relevant features of "conversation implication," as Grice would call it.[17] If S is asked, "But how do you know that those animals which you believe to be zebras are not cleverly disguised mules?", new information will be available to S: namely, S will now be aware that someone is questioning the adequacy of his/her evidence for the claim that the animals are zebras. S will wonder, legitimately, about what the questioner might know that he/she does not know. That is, after the zebras' authenticity has been challenged, S may no longer be justified in believing that the animals are not cleverly disguised mules. But note that S will also lose his/her justification for the claim that the animals are zebras. As Gail Stine put the point: "If someone questions whether I know the animals are not painted mules, he is suggesting that perhaps I didn't know that they were zebras."[18] Prior to the doubts having been raised, both p and q were justified; after the doubts have been raised, neither p nor q are justified. Thus, since whenever w_1 is "neutralized," p and q are also neutralized, this purported objection actually provides some evidence *for* the Contrary Consequence Elimination Principle rather than evidence *against* it.

There is one final possible objection to counting p as adequate confirming evidence for q which would be useful to examine here, not because it is more plausible than the arguments considered so far, but rather because it will clarify the distinction between the Contrary Consequence Elimination Principle and the Defeater Consequence Elimination Principle. That, in turn, will shed light on future tasks of this chapter; namely, presenting an agrument *for* the Contrary Consequence Elimination Principle and arguments against the Defeater Consequence Elimination Principle.

This objection is based upon another purported counterexample to the Contrary Consequence Elimination Principle. It might go as follows:[19] Suppose that I have an appointment to meet Ms. Reliable at 2:00 p.m. She calls me at 1:00 p.m. to confirm the appointment, and, knowing her to be a reliable person, I believe that she will keep the appointment. Suppose further that such a belief is justified. (Add whatever is required for confirming evidence.) Suppose also that if she were to receive a phone call from her landlord, Mr. Prosperous, in which he says that her apartment building has burned down, she would not keep the appointment. Thus, if the Contrary Consequence Elimination Principle is correct, I would be justified in believing that she will not receive such a phone call. But, surely, this objection would conclude, I am not entitled to believe that she will not get such a phone call. For what evidence do I have for that? The evidence which I have for the claim that Ms. Reliable will keep the appointment does not confirm that she will not receive the phone call from Mr. Prosperous, *and* it is not confirmed by the proposition that she will keep the appointment.

What are we to make of this objection? First, I think that it should be granted that I can be justified in believing that Ms. Reliable will keep the appointment and that I am not justified in believing that she will receive the phone call from Mr. Prosperous. But it is crucial to note that this is not really a counterexample to the Contrary Consequence Elimination Principle at all. The proposition *Ms. Reliable will receive a phone call from her landlord* is not a logical contrary of the proposition *Ms. Reliable will keep the appointment*. It may be true that if Ms. Reliable will keep the appointment, she will not have received such a phone call. But whatever the proper analysis of that conditional is, it certainly is not an entailment. She could receive a second call indicating that the first one was a bad joke; or she could discover in time to keep her appointment that although her building was damaged her apartment was not damaged at all, etc. Thus, this case does not contain a counterexample to the Contrary Consequence Elimination Principle since it is possible both that she will keep the appointment *and* that she will receive the phone call.

Nevertheless, the example is relevant to our consideration of the plausibility of scepticism. At the end of section 2.2, I indicated that there were four interpretations of the sceptic's Basic Epistemic Maxim that in order for S to be justified in believing that p, S must be justified in denying H (where H is either H_c or H_d). Since, under one interpretation, the sceptical hypothesis is not a contrary of p but rather a defeater of the justification of p, one of the principles to be considered is the Defeater Consequence Elimination Principle:

Defeater Consequence Elimination Principle   For any propositions, x and y, (necessarily) if y is a defeater of the justification of x, then $Jsx \rightarrow Js \sim y$.

Although the Appointment Case does not provide us with a counterexample to the Contrary Consequence Elimination Principle, it does *appear* to be a counterexample to the more general Defeater Consequence Elimination Principle. That is, the proposition that *Ms. Reliable will receive a phone call from Mr. Prosperous in which he claims that her apartment building burned down* is a defeater of the justification for *Ms. Reliable will keep the appointment.* It is important to note that the Defeater Consequence Elimination Principle is more general than the Contrary Consequence Elimination Principle because, although every contrary is a defeater, not every defeater is a contrary. In other words, the Defeater Consequence Elimination Principle entails the Contrary Consequence Elimination Principle; but the converse is not true. The Zebra Case and the Appointment Case illustrate that point.

Now, if the Appointment Case is a counterexample to the Defeater Consequence Elimination Principle, then there will be cases in which S's justification is defective, but in which S is not justified in believing that the proposition which makes it defective is false. Thus, there can be defeaters of a justification which S is not justified in denying. We will have to consider the Defeater Consequence Elimination Principle in detail later. At this point I will only assume that the Appointment Case provides no reason for rejecting the Contrary Consequence Elimination Principle. If that is correct, and if I have succeeded in showing that the Dretske-like objections to the Contrary Consequence Elimination Principle are not sound, then we can conclude that Direct Scepticism based upon that principle cannot be so easily discarded.

Nevertheless, we have not yet considered any argument for the Contrary Consequence Elimination Principle except to say that (at least to some) it appears intuitively plausible. Sceptics very rarely argue for it, and it is often taken for granted. However, that intuitive plausibility could be easily undercut if the principle conflicted with even more basic intuitions. A Moorean, for example, might argue that it is more certain that he/she knows that p than it is that the Contrary Consequence Elimination Principle is true. If, therefore, the adoption of that principle would lead to scepticism, then that principle ought to be rejected.

In order to increase the plausibility of Direct Scepticism and block this Moorean attempt at "refutation," it is necessary to consider whether there is a good argument *for* the Contrary Consequence Elimination Principle. I believe that there is a convincing argument for that principle. It is a rather simple argument once a partial characterization of justification has been presented; for the principle can be derived from that characterization. Unfortunately, that characterization is, itself, not as simple as one would hope, primarily because of the nontransitivity of confirmation discussed in this section. But since the overall purpose of this book is to refute scepticism, I will resist (as much as possible) the temptation to say more than what is necessary about justification.

The partial characterization of justification presented in the next two sections will not only provide us with an argument for the Contrary Consequence Elimination Principle, but it will also help to provide a basis for our analysis of the other three interpretations of the sceptic's Basic Epistemic Maxim and for the account of certainty developed in Chapter Three.

2.6 Clarification of the Contrary Consequence Elimination Principle and the Strategy to Argue for It

Up to this point, I have only considered the objections to the Contrary Consequence Elimination Principle. I pointed out that those objections were based upon the mistaken belief that if some proposition, say p, entails another proposition, say q, and p is justified for S on the basis of e, then q must be justified for S on the basis of e. I argued that the confirming evidence for q need not be the same evidence which confirmed p. Moreover, I claimed that the evidence available to S for q was p itself.

But to show that the objections to the Contrary Consequence Elimination Principle are not well founded does not provide any good reason for accepting that principle. That is one of the tasks of the next two sections. Roughly, I wish to argue that the Contrary Consequence Elimination Principle can be derived from a general model of justification which is intuitively plausible and acceptable to both sceptics and nonsceptics. In addition, the proposed model ought to be acceptable both to those who favor a foundationalist approach and those who favor a coherentist approach to justification, since the Evil Genius Argument is not aimed at any particular account of justification. Consequently, if the argument for the Contrary Consequence Elimination Principle depends upon a model of justification compatible with only foundationalism or only coherentism, then the plausibility of scepticism itself would become severely limited.[20]

The argument for the Contrary Consequence Elimination Principle has two main steps. First, as mentioned above, I wish to develop a partial characterization of justification which is generally acceptable. Second, I will argue that if the necessary conditions of S's having a justification, so characterized, are fulfilled for a proposition, say p, and p entails q, then the sufficient conditions of S's having a justification for q are fulfilled. I say that I will provide only a "partial" characterization of justification or a "model" of justification, because some of the important concepts in the proposed analysis of justification will remain relatively unanalyzed. I hope it will become clear that the soundness of the argument for the second main step does not depend upon any particular specification of these constitutive concepts beyond what will be said.

First, however, some ambiguities in the Transmissibility Principle must be removed. For on some readings it is clearly false. If *S is justified in believing that x* entails *S believes that x*, or if *S is justified in believing that y on the basis of x* entails *S actually used x to arrive at the belief that y*, then the principle is clearly false. For a person, S, may be justified in believing that x, and x may entail y, but, nevertheless, S may fail to be justified in believing that y if the latter entails that S believes that y and that S *in fact* used x to deduce y. S may be epistemically timid and not believe all that he/she is entitled to believe or may be epistemically misguided but lucky and may believe something to which he/she is entitled, but believe it for the "wrong reasons."

But that is not what is at stake here. Both defenders and opponents of the transmissibility principle agree that the issue is not whether S *does* believe that y or whether S "uses" x to get to y. Rather, the issue is whether S has adequate evidence "available" to entitle the belief that y. One way of resolving this difficulty is to add restrictions to the antecedent in the transmissibility principle, as both Gettier and Thalberg do. As shown above (see section 2.5), they formulate the principle which we are considering as follows: For any proposition x, if S is justified in believing that x, and x entails y, *and S deduces y from x and believes y as a result of this deduction*, then S is justified in believing that y. This does succeed in removing the difficulty, but I believe that there is another, preferable method. We could explicitly recognize that *S is justified in believing that x* does not entail *S believes that x* and specifically limit what is meant by 'S is justified in believing that x' to something roughly equivalent to *S is entitled to believe that x* or *S has available a justification for x.* This seems preferable to me, since the justification condition in the traditional analysis of knowledge is usually understood to be logically independent of the belief condition. The first method leads to a conflation of those two conditions. In addition, since the acceptability of the Contrary Consequence Elimination Principle depends upon whether S is *entitled* to believe that y (whenever x entails y and S has a justification for x) regardless of whether S exercises that right properly, choosing the weaker analysis of justification allows us to focus upon what is central in the dispute. To repeat: the issue here is whether S is warranted in believing that y, if S is warranted in believing that x, and x entails y. As mentioned above, the strategy is to demonstrate that the necessary conditions for being justified in believing that x, whatever they may be, are sufficient for being justified in believing that y, provided that x → y.

However, before proceeding to that demonstration, one other clarification of the task is required. I have already alluded to it in passing (see my section 2.5). In the terminology I have adopted, the problem can be put as follows: Consider some proposition, p, which is justified for S and which entails some further proposition, q, and imagine either that q contains some concepts which S does not understand or that the entailment between p and q is not at all obvious to S. For example, suppose that p is the conjunction of the five Euclidean postulates and q is a theorem with a rather high number. Would it be correct to claim that (necessarily) if p is justified for S, then q is justified for S?

Again, there are two alternatives available to us. Either we can narrow the scope of the Contrary Consequence Elimination Principle to only those propositions which are "obviously" entailed by x and for which S has the adequate concepts; or we can take the sense of justification inherent in the principle to include the cases in which S is entitled to believe a proposition even though S fails to have the ability to believe the proposition.

Although the second alternative is compatible with what we have said thus far, I think that it will simplify the argument for the Contrary Consequence Elimination Principle if we restrict it to those propositions for which S has the requisite concepts and to those entailments which are obvious to S. I will explain that alternative in a moment; but first let me comment on the alternative which I wish to reject.

Recall that I have just taken the expression 'x is justified for S' to mean that S is entitled to believe that x, rather than the stronger claim that S's belief that x is warranted. Only the latter implies that S believes that x. I argued that if the Contrary Consequence Elimination Principle committed us to claiming that S actually draws the inferences to which he/she is entitled, then it would surely be false. For we may be entitled to believe some proposition, x, and not believe x.

Now, we may fail to believe that x because we either lack the requisite concepts or fail to see the entailment between it and a proposition which we do believe. Thus, we could take the sense of justification inherent in the Contrary Consequence Elimination Principle to include cases in which S is entitled to believe that x but fails to do so because S lacks the ability to believe that x. This would simply be a special case of S's failing to believe some proposition which he/she is entitled to believe.

However, there are reasons for preferring the alternative way of accounting for the difficulty, namely, limiting the scope of the Contrary Consequence Elimination Principle to propositions which S is capable of deducing from those propositions for which he/she has a justification. First, in all the relevant cases, the entailed proposition is, in fact, understood by S, and the entailment is obvious to S. Specifically, in the Zebra Case, S understands that a cleverly disguised mule is not a zebra, and, as I have pointed out, in the Evil Genius Case, the proposition referring to the malevolent mechanism must be sufficiently understandable to S if it is to be one which the sceptic claims that S must be justified in denying. In fact, scepticism's plausibility depends upon formulating an immediately intelligible contrary to those propositions believed to be knowable—the contrary must be one which S and we understand is a contrary and it must be one which S and we are justified in believing to be false in order to have the knowledge we normally believe that we possess.

Moreover, one of the necessary desiderata of the model of justification which we will develop is that it delineates a concept of justification such that it and the other necessary conditions of knowledge are jointly sufficient for knowledge. Since one of those conditions is that S believes the proposition justified for him/her, we need only consider those propositions for which S has the requisite concepts. Hence, we will not be ignoring any candidates for knowledge if we include under the justification condition only those propositions for which S has the requisite concepts. For these reasons, I think that it is preferable to limit the scope of the Contrary Consequence Elimination Principle to those entailments which are obvious to S and to those propositions for which S has adequate concepts.

We are now ready to begin the argument for the Contrary Consequence Elimination Principle. To reiterate, the principle I will argue for is the stronger Principle of the Transmissibility of Justification Through Entailment. The Contrary Consequence Elimination Principle is a corollary of that stronger principle, since the denial of each contrary of a proposition is entailed by that proposition.

The argument in the next two sections is designed to show that if x entails y, and x is justified for S, then y is justified for S. I will take it that what is meant by the transmissibility principle is: If x entails y, and S is entitled to believe that x, then S is entitled to believe that y, where 'x' and 'y' range over contingent propositions understood by S and the entailment is an "obvious" one. (The reason for restricting the propositions to contingent ones will be explained later; see sections 2.7 and 2.8). I will provide a general argument for that principle by showing that if the necessary conditions obtain for x to be justified for S, and if x (obviously) entails y, then sufficient conditions obtain for y to be justified for S. But that requires developing at least a partial characterization of the general conditions of justification—the task of the next section.

2.7 A Partial Characterization of Justification: Confirming and Overriding Evidence

Roughly speaking, we can say that a proposition, x, is justified for S if and only if (1) S has available enough evidence, obtained in an epistemically reliable manner, which confirms x, and (2) S does not have evidence available which overrides that confirmation of x. Although such a characterization of jusification is so imprecise that as it stands it cannot assist us in determining whether the Principle of the Transmissibility of Justification Through Entailment or, for that matter, any other general principle of justification, is correct, still it can point us in the right direction. For it shows that in order to give a full characterization of justification we would need to explain carefully what is meant by *available evidence*, by *epistemically reliable*, by *confirmation*, and finally by *overriding evidence*; but, for our purposes, I believe that only the notions of *evidence available for S* and *overriding evidence* need be examined in detail. The other notions need to be explicated only insofar as they assist us in explaining those crucial concepts. That is the case because the truth of the transmissibility principle depends primarily upon the notions of available confirming evidence and available overriding evidence. The reasons for this should become apparent as the discussion progresses.

In order to avoid the appearance of developing an elaborate account of justification which merely leads to substantiating the sceptic's Basic Epistemic Maxim interpreted as the Contrary Consequence

Elimination Principle, I will point out that the analysis of justification has several other desirable features. For example, it can be shown that if x is justified for S, and y is justified for S, it does not follow that the conjunction, x&y, is justified for S. This result is especially useful in resolving the so-called Lottery Paradox.[21] In addition, we will be able to show that if the disjunction, $x \vee y$, is justified for S, it does not follow that either x is justified for S or y is justified for S. Now, of course, these results and the transmissibility principle would be expected if probability theory provided the semantics for the theory of justification; my account will not appeal to features of probability theory, however, primarily because such an appeal would limit the generality of the model.

Some of the results of the proposed partial characterization of justification are summarized below. If we let x, y, z range over contingent propositions understood by S and restrict the entailments to ones which are obvious to S, then the following principles result from the model:

(1)	Transmissibility Principle	If x entails y, and x is justified for S, then y is justified for S.
(2)	Conjunction Principle	If (x&y) is justified for S, then x is justified for S, and y is justified for S; but the converse is not true.
(3)	Disjunction Principle	If x is justified for S, or y is justified for S, then $x \vee y$ is justified for S; but the converse is not true.
(4)	Nontransitivity Principle	If x justifies y for S, and y justifies z for S, it does not follow that x justifies z for S.
(5)	Contradictory Exclusion Principle	If S is justified in believing that x, then S is not justified in believing that $\sim$x.

We can begin our partial characterization of justification by agreeing with Dretske that if S is justified in believing a proposition, x, then there is a set of propositions, w, *available to* S to use as an adequate reason for x. In other words, if S is entitled to believe that x, it will not suffice that there *is* adequate evidence to confirm x. Rather, that confirming evidence must be available to S—i.e., S must have (in senses to be defined) that confirming evidence.

But here we immediately encounter an important ambiguity which must be removed. For propositions can be "available" in two rather different ways—on the one hand, they can be propositions actually subscribed to or accepted by S, but, on the other hand, they can be propositions evidentially based or grounded on other actually accepted propositions. The former is *relatively* unproblematic; the latter requires a good deal of further explanation.

I say "relatively" unproblematic because propositions which are available in this first sense include at least two sorts: those inscribed in what have often been called "occurrent" beliefs as well as those inscribed in what have been called "dispositional" beliefs. I do not propose to say much about this first sense in which propositions can be available; for to do so would require that we examine many issues not relevant to our present purposes. It suffices to indicate that this category includes propositions which a person is consciously endorsing as well as those propositions which a person is disposed to endorse. Thus, I may not now be entertaining the proposition that my home is three miles from my office or the proposition that it is two miles from my office to the library, but those propositions are available to me in the first sense if they are needed as reasons for a further belief. I could call upon them to confirm my belief that it is farther from my office to my home than it is from my office to the library. I will call such propositions, inscribed in S's occurrent or dispositional beliefs, *propositions actually subscribed to by S*.

But, as I mentioned, the second way in which a proposition may be available does need further clarification. Suppose that S believes some proposition, say w, which is adequate evidence to confirm another proposition, say x; that is, wCx. S may fail to subscribe to x, but, nevertheless, I want to suggest that x is *available* for S. That is, x is available for S to use as a reason for further beliefs. Just as money once deposited earns interest, and that interest earns interest, etc., propositions may be said to compound and become available for S not because they were actually subscribed to, but rather because they are confirmed by propositions actually subscribed to.

An example or two may help to clarify what the second sense of *available evidence* means. Suppose that Miss Marple, the famous detective, is investigating a murder and has narrowed down the list of suspects to either Smith or Wesson. In addition, she believes that the poison used to commit the murder was purchased by the murderer on December 5 at the local drugstore. She discovers some plane tickets in Smith's desk and comes to believe that Smith was out of town from December 3 through December 7. Now, we would surely be amazed if Miss Marple failed to conclude that Wesson was the murderer, because she has available all the evidence which she needs. But notice that some of the evidence "available" for Miss Marple is not actually subscribed to by her or, at least, it *may* not actually be subscribed to by her. For example, the propositions that Smith was out of town on December 5, that Smith did not buy the poison, and, finally, that Smith is not the murderer are all necessary steps in determining that Wesson is the murderer. Those propositions may be available for Miss Marple only in the second sense; namely, the sense in which a proposition is available to her because it is confirmed by some proposition which she actually subscribes to or because it is confirmed by some proposition which is, itself, confirmed by some proposition to which she subscribes, etc. This is the sense in which propositions are available to Miss Marple to use as confirming evidence, even though she may not subscribe to them. The proposition that Smith is not the murderer is available because the proposition that Smith did not buy the poison is available, but neither of them may be subscribed to by Miss Marple. She may not conclude that Wesson is the murderer because she fails to make use of the evidence available to her; but the reason for lowering our estimation of her abilities as a supersleuth depends on the fact that she has all the needed evidence available.

In this case, many of the confirming steps are deductive. But that is not a necessary feature of the recursive character of availability. For example, suppose that I believe that Brown drives a Ford, garages it, has a valid-looking title to it, and claims that he bought the Ford two years ago when it was new. Let us assume that these propositions confirm the further proposition that Brown owns a two-year-old Ford. Suppose that I also believe that the Fords produced two years ago were fairly decent machines. The proposition that Brown's Ford is a fairly decent machine is surely partially confirmed. Perhaps we would want to add that Brown is a good driver and takes proper care of the Ford. The point is that the proposition that Brown owns a fairly decent machine is *available to* me to use to confirm further propositions. For example, it could be used to confirm a belief concerning the likelihood of arriving at a destination on time if I travel with Brown. That, in turn, could be used to confirm my belief that it is likely that I will arrive in time for the lecture, etc.

At the end of this section we will have to consider whether I have begged any issue against the sceptic by arguing for this characterization of available evidence. In particular, we will have to determine whether I have presupposed that some of our beliefs are justified or confirmed. And we will soon be refining the notion of availability. But I take it that these examples show that there are two senses in which propositions may be *available to S* to use as reasons for beliefs. The available propositions may themselves be propositions actually subscribed to by S, or they may be links of chains of propositions anchored in S's actually subscribed-to beliefs and joined by the confirming relationship. Let us say, then, that in order for a proposition to be available to S it must either be actually subscribed to by S (occurrently or dispositionally) or linked by the rules of confirmation directly or indirectly to those propositions actually subscribed to by S.

The following important questions remain to be answered: Are *all* propositions in such chains available for S to use as reasons to confirm beliefs? And are *all* actually subscribed-to beliefs available for S to use as reasons to confirm beliefs? We will turn to those shortly; but there is one issue which should be addressed immediately.

It may appear that I have begun the construction of the proposed model of justification in a way which excludes a coherentist account. Recall that I wanted the model of justification to be acceptable to both the coherentist and the foundationalist in order not to limit the plausibility of scepticism. So if this model were to bias the account in a way which made it unacceptable to the coherentist, it would already have encountered a serious difficulty. There is no foundationalist bias here, however, because I have not placed any restriction on the nature of the anchoring links. Some of them may be propositions which are not confirmed by other propositions subscribed to by S. Still others may be indirectly confirmed propositions, namely, propositions such that they would not be confirmed unless other propositions were confirmed. Thus far, I have made only the very minimal claim that if x is justified for S, then there must be some proposition, w, available to S which confirms x. I have begun to explicate what I mean by propositions which are available to S. Nothing requires or, for that matter, prohibits w from being identical to x. In addition, it could be the case that w confirms x and x confirms w. We will have to reconsider this question as the account develops, however, in order to determine whether at some point the account loses its neutrality.

Let us begin to answer the other questions concerning the extension of the propositions available for S's use as reasons. It may be thought that we could characterize this set as the set of *all* propositions subscribed to by S (either occurrently or dispositionally) and *all* the propositions related by rules of confirmation to those subscribed to by S. That is, it may be tempting to define the set of propositions available to S to serve as reasons—let us call them the *set of grounded propositions for S*—in the following recursive fashion: Suppose that S subscribed to some proposition, say p; thus, p would be available for S; and if p provided S with an adequtae confirming reason for q, q would be available to S; and if q . . . (etc.). This would require that we assume that all beliefs actually subscribed to by S are grounded for S and that a simple principle of recursion applies which extends the set of grounded propositions. I will return to that first assumption shortly, but let us consider the principle of recursion inherent in this characterization. Let us call the principle which embodies this simple recursive notion of groundlessness the *Simple Principle of Grounding*; it could be represented as follows:

Simple Principle of $(x)(y) [(Gsx \ \& \ xCy) \rightarrow Gsy]$
Grounding (SPG)

I wish that this principle were correct, for it would make the account of the model of justification and the argument for the transmissibility principle much more direct. But, unfortunately, the considerations which showed that the relation of confirmation was not transitive also serve to demonstrate that SPG is incorrect.[22]

Recall the case of Jones, the clever car thief. In that case there were three central propositions which we can abbreviate in this way:

w′   Jones is a clever car thief and has stolen a car. Clever car thieves almost invariably do those things which justify us in believing that they own the car that was stolen.

p′   Jones behaves in the way that car owners do.

q′   Jones owns a car.

In this example, it is clear that w′Cp′ and p′Cq′. But, as I argued earlier, it should be equally clear that $w'\mathcal{C}q'$. Let us assume, for the moment, that every belief actually subscribed to by S is grounded and that w' is actually subscribed to by S. There will be a chain of propositions, beginning with w' and ending with q'. The chain will look like this:

$$w'Cp'Cq' \ldots$$

Now, if the Simple Principle of Grounding were correct, q' would be available to S to use as a reason for further beliefs through such a chain. But surely S is not in a position to use q' as a reason for further beliefs, because the evidential ancestry of q' includes an overrider of the evidence for q'. That is, w' is *both* the evidential ancestor of q' and the overrider of p' for q'. For $(w'\&p')\mathcal{C}q'$. The chain has "degenerated" at q'. For neither q' nor anything following q' in the chain ought to be available to S, unless there are other nondegenerate chains including them. I will call those chains which do not degenerate either *nondegenerate chains* or *pure chains*. Of course, a chain may be pure up to a particular link; it may degenerate at that link and remain degenerate from then on. If all chains were pure chains, SPG would be correct. But the fact that some chains are degenerate requires the rejection of SPG.

In order to clarify the difficulty caused by the existence of degenerate chains, let me recall one of the specific areas of agreement between the sceptic and his/her opponent. In section 1.4, I pointed out that both parties to the dispute are willing to agree, at least initially, that if some proposition, say e, provides adequate evidence for another proposition, say p, it is not required that e entail p..Since e does not need to entail p, we can imagine a proposition, say o, which contains sufficiently strong counterevidence against p so that $(e\&o)\mathcal{C}p$. Let us call such propositions *overriders*. Thus, o is an overrider of the adequate evidence, e, for p.

In the Clever Car Thief Case, although $p'Cq'$, there is an overriding proposition, w', because $(p'\&w')\mathcal{C}q'$. The proposition, w', is an unusual overrider because it is also in the evidential ancestry of both p' and q'. I will call such unusual overriders *internal overriders*. Typically, an overrider of adequate evidence, e, will not be one of the evidential links in the chain preceding e; it will be an *external overrider* located on some chain which does not include e. We will have to consider these external overriders later, but the problem here is to develop an acceptable characterization of grounded propositions for S or the set of propositions available to S to use as reasons for further beliefs.

We have already looked at one problem in developing such an account—degenerate chains. There is another, equally serious, related difficulty with SPG. Suppose that the overrider, o, of the reasons, e, for p is part of a "conjunctive" belief of S. That is, suppose that the propositions actually subscribed to by S include (e&o), e, o. Since S believes that (e&o), there is the following degenerate chain:

$$(e\&o)CeCp \ . \ . \ .$$

The chain is degenerate because (e&o) is an internal overrider of the evidence, e, for p. Thus far, our analysis of the propositions grounded for S would be adequate, provided that we can give an acceptable characterization of what makes a chain degenerate, because we would not want p to be available to S to use as a reason for further beliefs. Nevertheless, the task of developing an account of the propositions grounded for S is further complicated by the fact that, in the case which we are now considering, S does have a pure chain beginning with an actual belief and terminating in p. The chain is: eCp. Thus, we cannot revise the Simple Principle of Grounding by merely requiring that all evidential chains be pure ones. Some restrictions on the first, or "anchoring," link in chains must also be developed. Put another way, the evidential ancestry of a proposition is crucial if we are to determine in what way that proposition can be used by S as a reason for further beliefs even if that proposition is among the beliefs actually subscribed to by S.

So the problem of defining groundedness becomes twofold:

(1)   How are we to restrict the set of S's actual beliefs which can serve as anchoring links of chains (whether pure or degenerate); and

(2)   How are we to characterize the link at which chains (whether properly anchored or not) become degenerate?

With a little more precision in our terminology, we will be able to resolve these two difficulties. In addition to those definitions already presented (see section 2.3), let us use the following:

xC*y   x is an evidential confirming ancestor of y.

x$\mathcal{C}$*y   x is not an evidential confirming ancestor of y.

B_S the set of propositions actually subscribed to by S—both occurrent and dispositional.

Γ_S a subset of B_S such that every member of Γ_S, say x, is a member of B_S and there is no other member of B_S, say z, such that zC^*x and $x\mathcal{C}^*z$.

I suspect that the reason for including the qualification that $x\mathcal{C}^*z$ in the characterization of Γ_S will not be obvious. In addition, the arguments for the Principle of the Transmissibility of Justification Through Entailment and the other welcomed principles of justification will not depend upon the addition of that requirement. Hence, some explanation for including it is necessary.

Recall that the model of justification and, in particular, the characterization of those propositions capable of anchoring confirmation chains should be acceptable to both the coherentists and foundationalists. If the Γ_S were *defined* so as to limit its membership to those propositions in B_S which have no other proposition in B_S in their confirmation ancestry, the model would lose the required neutrality. For the coherentist would claim that at least some, if not all, of the anchoring links are mutually confirming. Adding the requirement that $x\mathcal{C}^*z$ allows mutually confirming propositions, if there be any, to be included in B_S as well as those which have no other proposition in B_S in their confirmation ancestry. In other words, the addition of the requirement that $x\mathcal{C}^*z$ serves only to preserve the neutrality of the account of justification. It will not be needed in the arguments for the principles of justification developed in this and the following section.

By restricting chains which can make propositions available to S to those anchored in Γ_S rather than the larger set anchored by propositions in B_S, we can make significant progress in solving the first problem of arriving at an acceptable account of groundedness. We have seen that this problem arises, in part, because one conjunct of a belief can override the other conjunct. Since only the conjunction, and not the individual conjuncts, would be in Γ_S, there would be no chain beginning with either of the conjuncts.

Thus, the problems raised by conjunctive beliefs can be solved by restricting the anchoring propositions to those in the Γ_S set rather than the larger B_S set; but such a restriction has a further related advantage. Consider the general case of a degenerate chain, pCqCr, which is degenerate because $(p\&q)\mathcal{C}r$. (The Clever Car Thief Case is such a case.) Now, if both p and q were in B_S, then there would be a nondegenerate chain anchored in B_S terminating in r; namely, qCr. But since pCq and q$\mathcal{C}$p, only p would be in B_S. Hence, if we limit the anchoring set of propositions to those in B_S, r will not be grounded for S, because the only chain terminating in r will be 'pCqCr', and that is a degenerate chain.

Lest the suggestion to use Γ_S rather than B_S to anchor confirmation chains appear ad hoc, let me point out the following points in defense of it. First, if the conjunctive proposition does not contain an internal overrider, all the resulting chains will be pure ones, other things being equal. So S will "lose" no belief to which he/she is entitled. The same is true with regard to the general case of degenerate chains just considered. In that case only p is in Γ_S. However, q is not lost to S, since the chain pCq is not (yet) degenerate. In addition, any proposition, say t, such that qCt and such that (p&q)Ct will be available to S. Thus, any proposition on nondegenerate chains beginning with p would be available; and none of those on chains which degenerate will be available at or after the point at which the chain degenerates.

This point could be put another way. No proposition for which S has adequate confirming evidence will be lost to S as a result of this restriction (unless it occurs at or after the point at which a chain degenerates), since every proposition in B_S which is not in Γ_S will be a link in a chain beginning with a proposition in Γ_S.

Another advantage of using Γ_S rather than B_S to anchor confirmation chains may now be obvious. For suppose that B_S contains two propositions, p and q, such that pC $\sim$ r and qCr. At first glance, it might appear that since r and $\sim$ r are each supported by confirmation chains that neither could be justified for S.

But suppose, further, that p and q are links in an extreme form of a degenerate chain, pCqCr, such that not only does (p&q)$\mathcal{C}$r but pC $\sim$ r. The Clever Car Thief Case would contain such an extreme form of a degenerate chain if the propositions in it were changed to refer to a particular stolen car rather than just "a car." Presumably, in such a case, even though both p and q are in B_S and qCr, *only* ~r could be justified for S. For the only evidence for r is q; and p is both the only evidence for q and the evidence for ~r. Consequently, it is crucial to trace the evidential ancestry of all propositions in B_S in order to identify those propositions for which B_S actually provides evidence.

Now, of course, it could simply be claimed that B_S (as a whole) does not provide any evidential base for r and that B_S (as a whole) does provide such a base for ~r. And, no doubt, such a claim is true. However, an explanation of that claim would require tracing the evidential ancestry of both propositions. For the configuration of confirmation chains in B_S limits the set of propositions for which B_S really provides evidence. By restricting the propositions in B_S which can anchor confirmation chains to those in Γ_S, a general method has been developed which automatically organizes B_S into confirmation chains and provides a basis for identiying those propositions which could be justified for S.

In addition, this initial method of characterizing the anchoring propositions allows us to correctly capture the relevant intuitions in the following related case:

Let S's actual beliefs,
i.e., B_S, be as follows:

$$\left\{ \begin{array}{l} l \\ l\&m \\ l\&m\&n \end{array} \right\}$$

Let the following confirmation relationships obtain:

lCh
$(l\&m)$C ~ h
$(l\&m\&n)$Ch

In this case, a full characterization of B_S set shows that h is confirmed; while a partial characterization of B_S set confirms ~h. In such a case, it is clear that only h should be available to S to use as evidence for further beliefs. Now, if all the beliefs in B_S were permitted to anchor chains, both h and ~h would be available to serve as reasons for further beliefs. Although, as we will see shortly, there are occasions when both a proposition and its negation are grounded for s (but both are not *justified* for S), this is surely not one of them, because the evidence for ~h is contained in the evidence for h. Since 'l&m' is not in Γ_S (because there is another proposition in B_S which confirms it but which it does not confirm, namely, l&m&n), it cannot serve to anchor further propositions for S. Thus, since there is no pure chain anchored in Γ_S which includes ~h, the proposition ~h would not be available to S. In addition, as we will also see shortly, if ~h were grounded for S, S could not be justified in believing that h. But although other reasons may prevent S from being justified in believing that h, surely (l&m) ought not to provide S with such a reason.

For (l&m) is literally a part of S's total evidence relevant to h, and that full characterization confirms h.

The following two cases may help to clarify further the effect of using Γ_S rather than B_S as the anchoring set of propositions. Consider two cases: Case I—in which S believes that (p&q), p, q, where neither p nor q is an internal overrider of the other; Case II—in which S believes that (p'&q'), p', q', where q' is an overrider of the adequate reason, p', for some proposition, say r', but q' is not an overrider of the adequate reason, p', for some other proposition, say t'.

CASE I

B_S	Γ_S	Chains	Grounded Propositions for S
		pure chains	
p&q	p&q	(p&q)CpC...	p&q, p, q
p		(p&q)CqC...	
q			
		degenerate chains	
		none	

CASE II

B_S	Γ_S	Chains	Grounded Propositions for S
		pure chain	
p'&q'	p'&q'	(p'&q')Cp'Ct'	p'&q', p', q', t'
p'		[because	(but r' is not grounded)
q'		(p'&q')Ct']	
		degenerate chain	
		(p'&q')Cp'Cr'	
		[because	
		(p'&q')₵r']	

Figure 2.

I trust that, once these cases are studied, it will be clear that the relevant pretheoretical intuitions are captured by my proposal; for all the chains make propositions available to S except the degenerate one in Case II. In other words, only those propositions in chains with internal overriders are not available to S; all others are available through pure chains anchored in Γ_S.

We have given an initial answer to the first of the two questions concerning the extension of the propositions grounded for S. Our answer is that only those propositions in Γ_S can serve as such anchoring links. A revision in the characterization of the anchoring set will be required soon; but the task before us now is to make more precise the solution to the second problem — the proper characterization of the distinction between pure and degenerate chains. In order to do so, let me return once more to the Case of the Clever Car Thief. As we saw, w' provided S with an adequate reason for believing that p', and p' is an adequate reason for believing that q'. However, since w' is both the evidential ancestor of the evidence for q' and an overrider of that evidence for q', that last proposition is not available for S. The chain degenerated at q' because there was an internal overrider.

A diagram may help to clarify the nature of a typical degenerate chain. Let us consider a general case. Imagine a chain beginning with a proposition, say y_1, in Γ_S and terminating in some link, y_n. The chain could be represented as follows:

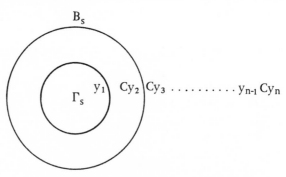

Figure 3.

As the diagram illustrates, some (or all) of the links following y_1 could be in B_S, but only y_1 can, by definition, be in Γ_S. Generalizing from this case, we can say that the chain '$y_1\,Cy_2\,Cy_3\,C \ldots\ldots y_{n-1}\,Cy_n$' degenerates at some link, say y_i, if y_i has an evidential ancestor, call it y_j, which is such that $(y_j \& y_{i-1})\,\mathcal{C}y_i$.

We are now in a position to give an initial definition of the set of propositions available to S to use as confirming propositions or propositions which can serve as reasons for further beliefs. A proposition, say x, is grounded for S (Gsx) iff either x is a member of Γ_S or there is a nondegenerate chain of propositions beginning with some proposition anchored in Γ_S and ending in x. A proposition is in Γ_S iff it is in B_S and there does not exist another proposition, z, in B_S such that z is in the ancestry of x, but x is not in the ancestry of z. These definitions could be put formally as follows:

$$Gsx =_{df} (x\epsilon\Gamma_S) \vee \{(\exists y_1, \ldots, \exists y_n) \ [(y_1\epsilon\Gamma_S) \& (y_1 Cy_2) \&$$
$$(y_2 Cy_3) \& \ldots (y_{n-1}Cy_n) \& (y_n = x)] \&$$
$$\sim (\exists y_i) (\exists y_j) (y_j \& y_{i-1} \mathcal{C} y_i)\}$$

where:
$$1 < i \leqslant n$$
$$j < (i-1) < i$$

$$x\epsilon\Gamma_S =_{df} (x\epsilon B_S) \& \sim (\exists z) \ [(z\epsilon B_S) \& (zC^*x) \& (x\mathcal{C}^*z)]$$

I said that this was an initial definition of groundedness, because some philosophers might want to add a further condition, namely, that the first or anchoring link in the chain of grounded beliefs must be obtained in an epistemically reliable manner. For, as it now stands, all of S's beliefs in Γ_S will be grounded beliefs for S. A belief in that set could have been acquired in a completely epistemically unreliable way. For example, it could have been acquired as a result of a posthypnotic suggestion or when under the influence of a belief-distorting drug. In addition, one might want to claim that any proposition grounded on an unreliably obtained belief is not available to S as a reason for further beliefs. Thus, one might take the notion of groundedness to include some requirement concerning the way in which the belief in question was acquired.

This seems to me to be an entirely legitimate requirement. The conditions in which beliefs can be acquired reliably are difficult if not impossible to specify in a general manner.[23] But whatever constitutes the conditions of reliable belief acquisition, this added requirement will make no difference to the argument for the transmissibility of justification as long as epistemic reliability is itself transmissible through the rules of confirmation in the way in which we have already specified groundedness is transmissible. That is, the same reasons which forced us to surrender the Simple Principle of Grounding will refute the suggestion that whenever x is reliably acquired and xCy, then y is reliably acquired; but we can respond with a similar strategy for revising the reliability condition.

But in order to accommodate this requirement fully, we will have to make a slight amendment in the characterization of the Γ_S set or the set of propositions which anchor propositions for S. To see this, suppose that B_S includes the propositions from the Clever Car Thief Case, or any set such that pCq and qCr but (p&q)₵r. Suppose further that p was acquired in an unreliable fashion, but that q was acquired in a reliable manner. If we were to define the Γ_S as previously suggested, p would be in Γ_S, but q would not be in that set, since pϵB$_S$ and pC*q but q₵*p. Consequently, the only chain anchored in Γ_S terminating in r would be a degenerate one, and r would not be grounded for S. This seems to me to be a counterintuitive result, because r ought to be available to serve as a reason for further beliefs, since r can be confirmed by q, and q was obtained reliably.

Let me point out that, even though r is grounded for S, it will not be justified for S. For reasons to be given shortly, I will argue that p will still be an overrider simply because it is in the Γ_S as previously defined. That is, I will argue that overriders need not be reliably obtained. But intuitions may vary here about a variety of things. Some may disagree with me because they believe that overriders must be reliably obtained. Thus, some may believe that r is justified, and others may not agree with my reason for believing that r is not justified. However, I trust that it will become apparent shortly that the argument for the Transmissibility Principle does not depend upon the particular specification of the model of justification consistent with these varying intuitions. The Transmissibility Principle follows from all of them! I wish merely to show how this model can be adjusted to accommodate these conflicting intuitions. For the sceptic's argument will then become that much more compelling because of its generality.

The minimal change in the procedure for specifying the propositions in the anchoring set of propositions for S must satisfy the following requirements: (1) it must select only those propositions which were obtained in an epistemically reliable manner and (2) it must not sacrifice the advantages of our original characterization of the Γ set for S—i.e., the problems related to "conjunctive beliefs" and degenerate chains must not re-emerge. The first requirement is easily satisfied. We shall merely restrict the propositions in the anchoring set to those in B_S which were reliably obtained. Let us call that set the R (reliably obtained) set for S, or more simply, the R_S.

But we must be careful not to include all the propositions in R_S in the anchoring set of propositions. For suppose that S has obtained p in a reliable fashion. Now, if pCq, and if epistemic reliability were transmissible through confirmation, then q would be obtained in a reliable fashion. Now, if qCr, the proposition r will be grounded for S. But if (p&q)Cr, the problems arising from the Clever Car Thief Case and conjunctive beliefs will re-emerge here.

A natural extension of the procedure adopted earlier will help us to solve these problems. Let us introduce the notion of *a belief's reliability depending essentially upon the reliability of another belief.* To clarify this idea, consider an S who has two beliefs, x and y, and suppose that S has obtained y in an epistemically reliable manner, and suppose that since yCx, the proposition x is reliable for S. Thus, both x and y are in R_S. Furthermore, suppose that x is reliable only because its reliability has been obtained through the transmission of reliability through confirmation from y. In this case, we can say that *x's reliability depends essentially upon y,* since yC*x, and x would not be reliable for S but for the reliability of y. Let us take the predicate 'xRy' to mean that yC*x and that x's reliability depends essentially upon y.

Let us now redefine the anchoring set of propositions for S, as follows: x is in the reliable anchoring set of beliefs for S, or Γ_S^R, if and only if x is in R_S and there does not exist another proposition, say y, in R_S such that xRy but yℝx. More formally:

$$x \epsilon \Gamma_S^R =_{df} x \epsilon R_S \,\&\, \sim(\exists y)\,(y \epsilon R_S \,\&\, xRy \,\&\, yℝx)$$

This is formally analogous to the previous definition of the Γ_S for S and satisfies the requirement that the anchoring beliefs be reliably obtained.

Since this is formally analogous to the previous characterization of the Γ_S, no change in the definition of groundedness is required except to substitute Γ_S^R for Γ_S. We will say that a proposition is grounded for S either if it is in the Γ_S^R or it is a link in a pure chain anchored in Γ_S^R.

I think that the only objection to this definition could be that it has the *apparently* counterintuitive result that there can be propositions which are justified for S but which are such that S's actual belief in them arose in an unreliable fashion. But this only has the appearance of a counterintuitive result. To see this, let us examine such a case. Suppose that B_S contains two propositions, p and q, and that there is a pure chain from p to q but not from q to p; that is, pC^*q but qC^*p. In addition, suppose that p was obtained in a reliable fashion but that the belief that q did not arise in a reliable manner. According to our definition of Γ_S^R, only p would be able to anchor propositions for S; but since there is a chain from p to q, q would be grounded for S.

Now, why should a proposition, say q, obtained in an unreliable fashion be available to S to use as a reason for further belief? And, other things being equal, why should such a proposition be justified for S? The answer depends upon noting that in such cases q is grounded for S not because it was in B_S, but rather because it is confirmed by a proposition obtained in a reliable way. Thus, other things being equal, q would be justified for S. That seems entirely correct, since what is required for S to be justified in believing that q is that S is entitled to believe that q. That entitlement depends upon q being linked to p; it does not depend upon the unreliable manner in which S obtained q which excluded q from Γ_S^R. To return to an analogy used earlier, if S has a valid title to a piece of property, S is entitled to the property, even though S may also have a forged title. Similarly, q is grounded through p, even though it also is in B_S in an unreliable manner.

Once again, it is important to point out here that although this formulation may again appear to incorporate a foundationalist bias, that is only an appearance. For this characterization of groundedness allows propositions to be grounded for S which form confirmation circles in Γ_S^R. Thus, if some proposition, say p, were in the ancestry of another, say q, and q was in the ancestry of p, neither p nor q would necessarily be excluded from Γ_S^R. Furthermore, suppose that both p and q were in R_S, and that pRq and qRp, then both could be in Γ_S^R. Presumably, the coherentist and foundationalist would differ over the extent to which propositions formed coancestral evidence-reliability chains; but the point is merely that the model is compatible with both accounts.

I will take it that we have completed the first step in developing a full enough account of justification in order to demonstrate that the sceptic is correct in claiming that if x entails y, and x is justified for S, then y is justified for S. Our account of justification is full enough if we can determine whether, whenever the necessary conditions of x being justified for S are fulfilled, the sufficient conditions of y being justified for S are fulfilled. One of the necessary conditions of x being justified for S is that there is a grounded proposition, w, for S such that w confirms x.

Grounding Condition
of Justification
$$(x)[Jsx \rightarrow (\quad w)(Gsw \,\& \, wCx)]$$

Before proceeding to the next major step — the explication of overriding propositions — a word or two must be said about confirmation, since it does play an important role in this necessary condition. It is, however, one of those constitutive concepts of justification which, as I mentioned earlier, will be left relatively unanalyzed.

It may appear as though I am begging the question against the sceptic by supposing that there may be adequate confirming evidence for propositions such as *there are zebras in the pen*. However, an analysis of what constitutes adequate confirming evidence is not necessary to resolve the issues at stake. The nature of adequate confirming non-deductive evidence *would* have to be determined before a complete characterization of justification could be developed. But, for our purposes, we can take 'xCy' to mean roughly that it is permissible according to the rules of evidence to infer y from x or that the truth of x is sufficient to warrant the truth of y. The only restrictions on the confirming relationship accepted thus far have been:

(1) The sceptic cannot insist that the confirming relationship be *limited* to entailments; that is, the sceptic cannot insist that if xCy, then $x \rightarrow y$ (see section 1.4).

(2) The confirmation relationship is not transitive (see the Clever Car Thief Case).

(3) If eCx and $x \rightarrow y$, it does not follow that eCy (see the Zebra Case in section 2.5).

(4) The analysis of the confirming relationship must be consistent with the claim that knowledge entails absolute certainty (see section 1.4).

A few additional restrictions will be developed shortly; for example, for contingent propositions, x and y:

(5) If $x \rightarrow y$, then xCy

(6) $(x \,\& \sim x) \, \mathcal{C} y$

(7) $(x \,\& \sim y) \, \mathcal{C} y$

Our present purpose, however, is to examine the sceptic's claim that if S is justified in believing that p, and p entails q, then S is justified in believing that q. I am suggesting that one of the necessary conditions of being justified in believing that p is that S has available some evidence, e, which confirms p either deductively or nondeductively. The sceptic would, no doubt, agree with that and would also believe that S never has adequate nondeductive evidence to confirm p. I am not supposing that S ever has that adequate confirming evidence for p nor for that matter, that S is ever justified in believing that p; I am merely indicating what would have to be the case *if* S were to be justified in believing that p. I take it that both the sceptic and his/her critic would agree that having adequate confirming evidence available for p is a necessary condition for being justified in believing that p. It is true that the nature of adequate nondeductive confirming evidence would have to be determined before a complete theory of justification could be developed. But that is not the task at hand. Here the issue is simply this: Is the sceptic correct in claiming that whenever S is justified in believing that p, and p → q, S is justified in believing that q? Put in our terminology: *If* S has adequate nonoverridden confirming evidence available for p, and p entails q, does S have adequate nonoverridden confirming evidence available for q? In order to decide the answer to that question, we do not need to presuppose that S has adequate confirming evidence available for p.

Nevertheless, there is one point worth mentioning about confirming evidence, and raising it will assist us in contrasting confirming evidence with overriding evidence. Suppose that a conjunction, say (p&q), confirms another proposition, say r. Consider two persons: S_1, who actually subscribes to p and actually subscribes to q, but has not "put p and q together," so to speak and, consequently, does not actually subscribe to (p&q); another person, S_2, who actually subscribes to the conjunction as well as to each conjunct. In our terminology, B_{S_1} contains p and it contains q, but does not contain (p&q); whereas B_{S_2} contains p, it contains q, and it contains (p&q). Clearly, other things being equal, S_2 has a grounded proposition which confirms r, since S_2 does believe that (p&q). But does S_1 have confirming evidence available for r? After all, S_1 has no belief which confirms r. S_1 does have the "makings" of such a confirming reason, because S_1 does believe that p and does believe that q. But it does seem that

S_2 is in a better epistemic position with regard to r than is S_1. (One is more nearly ready to bake a cake after the ingredients have been put together.) Recall the Smith and Wesson Murder Case. If Marple has not put together the evidence which confirms that Wesson is the murderer, she is less entitled to believe that Wesson is the murderer than she would be were she to "put two and two together."

I would hold that the confirmation chains must begin with propositions in Γ_S^R, not with a conjunction of propositions in Γ_S^R. This accords with the sceptic's basic intuition that it is difficult (if not impossible) for S to be justified, completely justified, in believing any proposition. In fact, the sceptic could legitimately require that in order for x to be justified for S, S must be in the best epistemic position with regard to x. I agreed at the outset of our investigation of scepticism to grant to the sceptic everything that could be reasonably granted. So let us grant that confirmation chains must begin only with those propositions in Γ_S^R.

It may seem that this is an odd place to make such a concession to the sceptic, because in this section I am arguing *for* the sceptic's Contrary Consequence Elimination Principle. But such a concession does not affect the soundness of my argument for the sceptic's position. It will be seen that the argument for the Contrary Consequence Elimination Principle goes through, if it does at all, regardless of the way in which this issue is resolved. This shows, once again, the centrality of that principle as well as of the Transmissibility Principle in an acceptable general analysis of justification.

We may now turn to the next step in the development of the general model of justification, namely, overriding evidence. The issue just discussed concerning whether confirmation chains are anchored by individual propositions instead of conjunctions of propositions in Γ_S^R will have to be reconsidered with regard to chains leading to overriders, but first I would like to make a few general comments about overriders.

A proposition, say u, is an overrider of the confirming evidence, e, for p if $(e \& u)\mathcal{C}p$. It should be obvious that there can always be an overriding proposition of the confirming evidence, e, for p as long as e fails to entail p. But even when e entails p, there can be overriding propositions, namely $\sim e$ and $\sim p$. That is, if eCp, then $(e \& \sim e)\mathcal{C}p$ and $(e \& \sim p)\mathcal{C}p$ regardless of whether e entails p. The reason for this is simply that evidence which is logically inconsistent with a proposition cannot confirm that proposition. Note that since, for most S's, the Γ_S^R will contain contradictory propositions, either we *must* restrict the anchoring links of confirming chains to propositions in Γ_S^R instead of conjunctions of propositions in Γ_S^R or we must grant that contradictory propositions are not confirming evidence. Otherwise, all propositions will be confirmed for S. Because of this, later we will have to restrict the range of propositions covered by the Transmissibility Principle to contingent propositions.

Since there are potential overriders of all confirming evidence, it should be apparent that even if S has grounded, adequate confirming evidence for x, S may not be entitled to believe that x. For S may also have other grounded beliefs, say u, which override his/her evidence for x. I take it that in such a case S would not be entitled to believe that x. That is, S may not be justified in believing that x even though S has adequate, grounded confirming evidence for x.

For example, one could have grounded beliefs for and against a proposition, say p. Return to the Zebra Case. S may have the evidence for p mentioned in Dretske's original presentation of the case. S may "know what zebras look like," and "the animals may be in a pen clearly marked 'Zebras.'" And, thus, S may have adequate confirming evidence (according to Dretske) for the proposition that the animals are zebras. But S may also have evidence against it based upon knowledge of the zookeeper's past painting activities. That is, S may know that the zookeeper has often painted mules to look like zebras. In that case, S would not be justified in believing that the animals are zebras, because S has reasons which override the evidence for that belief. Consider another case. There could be conflicting eyewitness reports of an automobile accident—one report grounding the proposition that A was responsible for the accident and another report grounding the denial of that proposition. But, as in the Zebra Case, the counterevidence need not be *that* strong to override; it could merely be such that it sheds sufficient doubt on p to bring it about that S is no longer entitled to believe that p.

We have already considered one class of *grounded*, overriding propositions—the internal overriders which prevent other propositions on the same confirming chain from becoming grounded. But external overriders, although not blocking propositions from being grounded, do block propositions from being justified for S. We have said that in general an overriding proposition is one which is such that it blocks or overrides adequate confirming evidence. Thus, if eCp, and o is an overrider of the confirming evidence, e, for p, then (o&e)$\mathbb{C}$p. If o is in the ancestry of e, i.e., if oC*e, then o is an internal overrider. But what about propositions which are not internal overriders?

It is crucial to note some essential differences or asymmetries between confirming evidence and overriding evidence. As we have said, the sceptic will, presumably, wish to make it as difficult as possible for S to have adequate *confirming* evidence for p. For example, he/she will insist that e must be in the Γ_S^R or on a pure chain anchored in Γ_S^R; that is, e will not be allowed to be a conjunction constructed out of grounded propositions for S. For similar reasons, the sceptic will want to make it relatively easy for S's confirming evidence to be overridden. That is, the sceptic will not require that the overriders be reliably obtained and will allow overriders to be conjunctions constructed from grounded propositions.

In order to understand the asymmetry between overriding and confirming evidence, let us return once again to the Clever Car Thief Case. Suppose that w' was obtained in an epistemically unreliable fashion and that p' was obtained in a reliable fashion. We argued that the sceptic will not permit w' to anchor confirming chains; but, no doubt, the sceptic will insist that the mere presence of w' in B_S is sufficient to block S from being justified in believing that q'. Since we can suppose that S does not know that w' was obtained in an unreliable fashion, it can serve to override the evidence, p', for q'. S would not be entitled to believe that q' until S had no overriding reason, whether grounded or not.

Consider another case, one which does not involve internal overriders. Suppose that Ms. Detective is a police officer trying to determine which of two persons was responsible for an automobile accident. All the evidence points to Mr. Imprudent. The skid marks indicate that Imprudent's car was traveling too fast. The brakes on Imprudent's car are not working properly. Suppose that Ms. Detective acquires these propositions in a reliable fashion. But suppose further that a purported eyewitness says that the accident was not Mr. Imprudent's fault at all and that Miss Innocent was speeding, ran a light, and somehow faked the skid marks. Let the eyewitness be acting under a posthypnotic suggestion caused by Mr. Imprudent; or let the eyewitness be Imprudent's mother-in-law. Again, it is not necessary to develop a characterization of the features of reliably an and unreliably obtained beliefs. Just let Ms. Detective acquire the proposition in an unreliable fashion, whatever that may be, that Miss Innocent was speeding, ran a light, and faked the evidence. The point is that although such a proposition would not be in Γ_S^R because it was obtained unreliably, it is, nevertheless, overriding evidence.

There is a second feature of the asymmetry between overriding and confirming propositions concerning epistemic reliability which must be noted here. Suppose that a proposition, say p, is in B_S but not in Γ_S^R or attached to a proposition in Γ_S^R by a pure chain. The proposition p is not grounded for S—i.e., it is not available as confirming evidence for further beliefs. The proposition p will be available as overriding evidence because it is in B_S, but in addition it would *seem* that any proposition attached to p by a pure chain ought to be available to S as an overriding proposition for the same reasons that any proposition in B_S ought to be available as overriding evidence.

In a moment I will explain why we cannot allow *every* proposition in B_S to anchor *chains* of overriders, but at this point it may be wondered why it is necessary to count *every* proposition in B_S as an overriding proposition, since, as we have seen, if a proposition is in B_S but not in Γ_S, it will appear on a chain anchored by a proposition in Γ_S. That is, what is gained by including all propositions in B_S as potential overriders, since any proposition not in Γ_S but in B_S will reappear on a chain anchored by Γ_S? The answer is that there may be a proposition, say r, which is in B_S but not in Γ_S and such that the only chain leading to it anchored by a proposition in Γ_S, say p, is a degenerate chain. In that case, if we did not allow every proposition in B_S to serve as a potential overrider, r would not be available to S as an overrider. That result is unacceptable since r may be in B_S but not because p (or another proposition) is in B_S. Recall the standard case of a degenerate chain: pCqCr and (p&q)𝒞r. Now suppose further that although qCr, S subscribes to r but not because S subscribes to q. In fact, suppose that S subscribes to r but not because S subscribes to any other proposition whatsoever. The coherentist would, no doubt, deny that there are any such propositions, but since our account is to be acceptable to the foundationalist as well as the coherentist, we must allow for that possibility. If there is such a case, we would want r to be available as a potential overrider merely because it is in B_S for the reasons already considered. Thus we cannot rely upon potential overriders re-emerging on pure chains anchored in Γ_S and we must count all propositions in B_S as potential overriders. Let me note in passing that a similar problem does not arise with confirming evidence. That is, no potential confirming evidence is lost by restricting such evidence to propositions in Γ_S^R and those propositions on pure chains anchored by propositions in Γ_S^R. In the case just considered, if r's reliability did not depend essentially upon p's reliability, r would be in Γ_S^R by definition.

Thus far we have seen two important differences between overriding and confirming evidence. First, any proposition in B_S is available as an overrider, and, second, any proposition in Γ_S or on a pure chain anchored by a proposition in Γ_S is available as an overrider. There is a third important asymmetry between overriding and confirming evidence which has already been mentioned. Presumably the sceptic would say that if S has the makings of an overriding proposition of e for p, S is not entitled to believe that p. To see this, suppose S has adequate confirming evidence for the claim that a close friend, Mr. Prevaricator, has never lied to him/her. Suppose further that, at a party, S overhears Mr. Prevaricator claiming that he delights in fooling a close friend of his whom he first met at a New Year's Eve party. In addition, S believes that he/she first met Mr. Prevariactor at a New Year's Eve party, but S does not put two and two together. That is, S fails to believe the conjunction 'Mr. Prevaricator lies to a person whom he met at a New Year's Eve party and Mr. Prevaricator met me (S) at a New Year's Eve party'. Hence, although the proposition that Prevaricator may have lied to S will not be grounded for S (because S has not put the confirming evidence together), that proposition can serve to override the confirming evidence for the proposition that Prevaricator has never lied to S.

Thus, in line with our policy of granting whatever can be reasonably granted to the sceptic, let us acknowledge these asymmetries between confirming and overriding evidence. The requirements for the former are stringent; for the latter they are relaxed. Once again, however, it will be apparent shortly that the soundness of the argument for the transmissibility principle does not depend upon whether we grant the asymmetry of overriding and confirming evidence.

Our discussion of overriding evidence up to this point could be summarized as follows: We will allow conjunctions of propositions in B_S to be overriders and we will allow conjunctions each conjunct of which either is in B_S or is a link of a pure chain which is anchored by a conjunction of propositions each of which is in Γ_S. It would be incorrect to allow every proposition or every conjunction of propositions which are in B_S but not in either Γ_S or Γ_S^R to anchor chains, because that would allow us to ignore the evidential ancestry of a proposition. Some propositions which ought not to be available as overriders would become available. To see this, reconsider the standard degenerate chain: pCqCr and (p&q)$\mathcal{C}$r. Suppose that p and q are in B_S (and r is not in B_S) but S believes that q only because q is confirmed by p. In such a case, r ought not to be available as an overrider; but it would be if we allowed every proposition in B_S which is not in either Γ_S or Γ_S^R to anchor chains, because q would be allowed to anchor a chain and the chain 'qCr' is a pure chain.

We must be careful here not to exclude some genuine potential overriders. Again, consider the standard degenerate chain: pCqCr and (p&q)$\mathcal{C}$r. This time, suppose that S believes that q but not because S subscribes to p. In fact, suppose that S subscribes to q but not because S subscribes to any other proposition. (Again, the coherentist would demur, but we must allow for the foundationalist view.) The proposition, q, is not in Γ_S because pCq and q$\mathcal{C}$p. In addition, since we can assume that q is not in R_S, it will not be in Γ_S^R. However, in this case it does seem as though r ought to be available as an overrider, because its actual evidential ancestry begins with q and not with p. Thus, we will have to make some provision for propositions in B_S that are not in Γ_S or Γ_S^R which are subscribed to by S but not because S subscribes to any other proposition. Let us call such propositions *autonomous propositions* for S, or A_S. We must allow such propositions to anchor chains of overriding evidence. Thus we must expand our class of overriding propositions as follows: Potential overriders will be conjunctions of propositions each conjunct of which is either in B_S or a link of a pure chain anchored by a conjunction of propositions each of which is in Γ_S or Γ_S^R or A_S.

There is one further point which must be noted before we can give a complete characterization of overriding evidence. Suppose that there is a nondegenerate confirming chain beginning with a proposition, y_1, which is anchored in Γ_S^R, and assume that the chain terminates in some proposition, say y_n. The proposition, y_n, is grounded for S; and if there is no overrider of the evidence for it, y_n will be justified for S. The chain will look like this:

$$y_1 Cy_2 Cy_3 Cy_4 \text{ .... } y_{n-1}Cy_n$$

Now, suppose further that there is a proposition, u, which is available as an overrider for S, such that $(u\&y_3)\mathcal{C}y_4$. It may also be that $(u\&y_{n-1})\mathcal{C}y_n$. In that case, y_n is clearly not justified for S. For u would be a straightforward external overrider of the confirming evidence, y_{n-1}, for y_n. But it could also be that $(u\&y_{n-1})Cy_n$. In that case, it would still seem that u prevents y_n from being justified for S, because u is an overrider of an evidential ancestor, y_3, of y_n. Although u is not a direct overrider of the confirming evidence, y_{n-1}, for y_n, it does override the confirmed proposition y_n, for S. It is an overrider of an evidential ancestor of y_n and, so to speak, pulls the props (or propositions!) out from under y_n. Thus, for the same reason that a proposition is not grounded for S if there is an internal overrider of an evidential ancestor of the proposition, it is not justified for S if there is an external overrider of an evidential ancestor of the proposition.

We are now ready to characterize an overriding proposition of a confirmed proposition for S:

> *u is an overrider of the confirmed proposition, x, for S* [Osux] iff (1) x is confirmed for S and u conjoined with an evidential ancestor, y_i, of x fails to confirm y_{i+1} *and* (2) u is a conjunction of propositions each of which is either in B_S or a link of a pure chain anchored by a conjunction of propositions each of which is in Γ_S or Γ_S^R or A_S.

Let us call propositions which satisfy the necessary conditions of overriding evidence but not the necessary conditions of confirming evidence *pseudogrounded propositions for S.* They are like grounded propositions because they are *available* for S to use, but they are *pseudo*grounded because they are *not* available to use as confirming propositions. They are only available for S to use as overriding evidence.

The second major task of this section is complete. For we have now characterized overriding evidence, and I believe that we are now, at last, in a position to formulate a general model of justification with which to test the Transmissibility Principle and its corollary— the Contrary Consequence Elimination Principle. I suggest, as a general model of justification acceptable to the sceptic and his/her critic, that *x is justified for S* iff S has a grounded proposition, w, such that w confirms x and that there does not exist a proposition, u, such that u is an overrider of x for S. More formally, the suggested model is:

Model of Justification $\quad$ (x)$\Big\{$Jsx ↔ (∃w) [Gsw & wCx &∼(∃u)(Osux)]$\Big\}$
(MJ)

I began this section with an intuitive characterization of justification. This proposed model differs from it only in the sense that two of the constitutive concepts—confirming evidence available to S and overriding evidence available to S—have been made sufficiently precise so that it is now possible to test the validity of the transmissibility principle and the other "welcomed" principles of justification mentioned at the beginning of this section. That is the task of the next section.

2.8 The Argument for the Contrary Consequence Elimination Principle and Other Welcomed Principles of Justification

We are in a position to indicate why the sceptic's Contrary Consequence Elimination Principle is correct. That principle asserts that if x and y are contraries, and S is justified in believing that x, then S is justified in believing that ∼ y. That principle is a corollary of the more general principle which we have called the Principle of Transmissibility of Justification, namely, if x entails y, and S is justified in believing that x, then S is justified in believing that y. The Contrary Consequence Elimination Principle is a corollary of the transmissibility principle because the denial of every contrary of a proposition is entailed by that proposition. I will argue for the Contrary Consequence Elimination Principle by showing that the stronger transmissibility principle is correct (with the restrictions indicated in section 2.6, plus one to be introduced shortly).

Given the Model of Justification (MJ) developed in the previous section, we can show that the transmissibility principle is correct if we can show that S has adequate, grounded, nonoverridden evidence for some proposition, say q, whenever S has a justificaion for some proposition, say p, and p entails q. Recall that my suggestion in replying to Dretske was that he had not provided any reason for eliminating p as the candidate for the adequate, grounded, nonoverriden evidence for q. I think I can now show that p is such a reason for q, and, consequently, that S has a justification for q.

The task is to show that p and q satisfy the three following conditions if p entails q, and p is justified for S:

(1)   p is grounded for S, i.e., Gsp.

(2)   p confirms q, i.e., pCq.

(3)   There is no overriding proposition, u, of the evidence, p, for q for S, i.e., $\sim (\exists u)(0suq)$.

For if (1) and (2) and (3) are true, then Jsq (by MJ).

We are given that the antecedent in the transmissibility principle obtains, i.e., that p is justified for S, and that p entails q. The task is to show that the sufficient conditions of q being justified for S are fulfilled. Now, if p is justified for S, then by MJ there must be a grounded proposition, w, for S such that wCp, and there cannot be an overriding proposition, u, of the evidence, w, for p. That is all that we are entitled to assume.

Nevertheless, that is sufficient to guarantee that p and q satisfy the three conditions required for q to be justified, mentioned above. That is, we can show that p is grounded, that it confirms q, and that there is no overrider of the evidence, p, for q for S.

It can be shown that any proposition is grounded for S, if it is justified for S. The proposition, p, is grounded for S if there is a *pure* chain anchored in Γ_S^R which terminates in p. We can show that there is such a pure chain if we can show that there is a nondegenerate chain anchored in Γ_S^R terminating in p. Now, since p is justified for S (by hypothesis), there must, by MJ, be a grounded proposition, w, which confirms p. That, in turn, implies either that $w \epsilon \Gamma_S^R$ or that there is a pure chain beginning with some proposition in Γ_S^R which ends with w. If $w \epsilon \Gamma_S^R$ and wCp, then there is a chain anchored in Γ_S^R ending with p, namely, the short chain wCp. On the other hand, if there is a pure chain beginning with some proposition in Γ_S^R other than w which terminates in w, then there is a chain from that other proposition to p which includes w as the link immediately prior to p. Thus, if p justified for S, then there is a chain anchored in Γ_S^R terminating in p.

But since the Simple Principle of Grounding (SPG) is not correct, the existence of such a chain is not sufficient to insure that p is grounded. For, after all, it might be thought that the chain could degenerate at p—given the above information about p. It cannot degenerate prior to p, because then w would not be grounded. But, perhaps, it could degenerate at p. That is what I meant when I said earlier that if SPG were correct, the argument in this section could be much more direct.

Nevertheless, it is clear that the chain cannot degenerate at or prior to p. The chain cannot degenerate at w or any point prior to w, because w is grounded for S. And it cannot degenerate at p. For if it did there would be a grounded proposition (one of the propositions in the chain prior to w) which when conjoined with w failed to confirm p. In other words, there would have to be an *internal* overrider of the confirmed proposition, p, for S. Thus, there would have to be a grounded proposition, u, such that $(u\&w)\mathbb{C}p$. But if p is justified for S, as per our hypothesis, then, by MJ, there is no such u—whether grounded or pseudogrounded. That is, since p is justified for S, there is no overrider of p for S of *any* sort; specifically there is no internal grounded overrider. Thus, the first condition is fulfilled. The proposition, p, is grounded for S.

Note that this argument for the claim that p is grounded for S is independent of the way in which we choose to particularize the general model of justification. The set of grounded propositions for S will vary depending upon the way in which we choose to characterize the set of propositions in B_S permitted to anchor chains for S. That set could be Γ_S or Γ_S^R or some other set. But *whatever* the composition of the anchoring set, if p is justified for S, there will be a nondegenerate chain from that anchoring set to w. That chain continues to p and it cannot degenerate at p because, since p is justified for S, there is no u available for S such that $(u\&w)\mathbb{C}p$. Since any proposition on the chain prior to w would be available to S, there is no such u on the chain, however it is anchored, prior to p. Even if chains were permitted to be anchored by an unreliably obtained proposition, there would still be a nondegenerate chain terminating in p. In addition, if overriders included only the grounded propositions for S and not the larger class of both grounded and pseudogrounded propositions, there would still be no grounded internal overrider of the evidence, w, for p, since p is justified for S. For, again, if p is justified for S, then according to MJ, there is no overrider of w for p of any sort, whether internal or external. So even if overriders were limited to grounded propositions for S, there would still be no internal grounded proposition. Similarly, if we were to limit overriding propositions to those already "put together" by S or those on individual nondegenerate chains anchored in Γ_S or Γ_S^R or A_S, then there would still be no internal overrider of w for p, because p is justified, and there are no overriders of any sort. Thus, the first of the three conditions sufficient to show that q is justified for S is fulfilled—p is grounded for S.

The second step in the argument to show that q is justified for S is the easiest step. We must show that pCq. But, as I argued in section 2.5, p is the best possible confirming reason for q, since it entails q. The general rule that if x entails y, xCy will have to be amended shortly because, for example, x could be a self-contradiction, and it does seem that in general inconsistent evidence cannot be used to confirm anything. I will propose that we restrict the range of x and y to contingent propositions.

One objection, however, to the general principle that if x entails y, then xCy, even with the restriction just mentioned, should be considered here. For it may appear that I am somehow begging the question against those who deny the validity of the Transmissibility Principle by claiming that (with the appropriate restrictions) if x → y, then xCy.[24] For example, if Dretske believes that the transmissibility principle is invalid, why should he be willing to grant that if x entails y, then xCy?

Of course, it cannot be true that my argument for the transmissibility principle is sound and that the transmissibility principle is invalid. And, thus, those who deny the validity of the transmissibility principle may wish to reject the claim that if x → y, then xCy. But the issue is: Could there be any good reasons for rejecting this claim, and are they the reasons which were initially advanced for rejecting the transmissibility principle? To put it another way, would the reasons for rejecting the transmissibility principle apply equally well to rejecting the claim that if x entails y, then xCy?

It seems that there is at least a *conceivable* reason for rejecting this particular way of stating what I intend to be claiming when I assert ⌜if x entails y, then xCy⌝. Suppose that we were (somehow!) to discover that our notion of an entailment did not correctly portray an inference that was necessarily truth-preserving. For example, suppose that after careful study we believed that some logics were unsound which contained an operator which we thought was the formal analogue of entailment (i.e., transitive, reflexive, and nonsymmetric) and that we could not identify any *particular* "suspect" feature of the logic. That *might* give us some reason, however slight, to doubt that our notion of "entailment" correctly expressed the necessary truth-preserving property intended to be referred to in 'if x entails y, then xCy'.

But that doubt, even were it possible, is based upon considerations which are irrelevant to the dispute between the sceptic and his/her critic concerning the validity of the transmissibility principle. I take it that the expressions "entails" and " → " in 'x entails y' and 'x → y' are meant to stand for a relationship which is such that it is necessarily truth-preserving. Thus, the transmissibility principle which we are examining expresses something like the following: If S is justified in believing that x, and necessarily, if x is true, then y is true, then S is justified in believing that y. It could be (on the fanciful hypothesis considered in the preceding paragraph) that our concept of this necessarily truth-preserving inference is flawed. But the issue is not whether there *are* any pairs of propositions (x, y) such that necessarily if x is true, y is true; or whether any particular concept of ours (e.g., entailment) accurately captures the extension of those pairs, but rather whether, if there is such a relationship between x and y, and if S is justified in believing that x, then S is justified in believing that y. If there are no such pairs of propositions, then the transmissibility principle becomes epistemically useless because nothing fulfills the antecedent; but it does not, thereby, become false.

Thus, what is expressed by the principle 'if x entails y, then xCy' is: For every pair of propositions (x, y), if, necessarily, x is true, then y is true, then x confirms y. That principle is analytically true since whether a proposition, say x, confirms another one, say y, depends upon whether x, if true, is a sufficiently good guarantee of y. Recall that 'xCy' means that, according to the rules of confirmation, it is permissible to infer y from x. A necessarily truth-preserving inference must be permissible. Thus, even if our concept used to portray those pairs of propositions, if any, were flawed, the principle which is expressed by 'if x → y, then xCy' is valid.

Furthermore, it is important to recall that the objection brought against the transmissibility principle was not based upon the rejection of the general rule I am proposing, namely: If x → y, then xCy. Rather, the objection was based on the claim that even though x entails y, and S is justified in believing x for some reasons, r, S may not be justified in believing that y for the same reasons, r. That is, the opponents of the transmissibility principle believed that its validity depended upon the truth of another related principle; in our terminology: If x → y, and eCx, then eCy. I granted that this related principle was false (because of the Zebra Case), but claimed that the transmissibility principle does not depend upon its truth. I shall not repeat that argument here. The point is that the objection to the transmissibility principle was not based upon the claim that if x entails y, x may not be sufficiently good evidence for y—but rather that, if x entails y, *the evidence* for x may not be sufficiently good evidence for y.

Thus, I will assume that (except for the restriction to be explained shortly) both the sceptic and his/her critics would agree that if x entails y, then xCy. And if that is true, then the second step in the argument for the transmissibility principle has been accomplished— p confirms q.

We have only to show that the third necessary condition of S's being justified in believing that q obtains. Now, since we have seen that it is possible that there is overriding evidence even of confirming evidence which entails p, what remains to be shown is that there is no overriding proposition, u, of the evidence, p, for q. Roughly, the reason that there is no such proposition is that, since p → q, any proposition that overrides q will override p. And since S is justified in believing that p, there can be no overriding proposition of any sort for p for S.

To see this, imagine what u would have to be in order to override the evidence for q, when that evidence is p, and p is such that it entails q. That is, what propositions would be sufficiently strong to override the evidence, p, for q? It could not be something that would normally be inductive counterevidence against q, because p entails q, and S has the grounded evidence p. It would have to be something strong enough to override an entailment. As I have mentioned earlier, if p entails q, the only possible overriding propositions of p for q would be either ∼ p or ∼ q. But if ∼ q is a grounded proposition or pseudogrounded proposition for S, so would ∼ p be a grounded or pseudogrounded proposition for S because, by hypothesis, p entails q, and, hence, ∼ q entails ∼ p. Consequently, ∼ qC ∼ p, and if ∼ q is available as an overrider for S, so is ∼ p. Thus, the only possible candidate for an overrider of the confirmation of q by p which we need to consider is ∼ p. But since we have seen that nothing in general prevents a proposition and its denial from being grounded, we must consider the possibility that ∼ p is grounded for S.

Well, can ∼ p be a grounded or pseudogrounded overrider for S? The answer is that it cannot be, for, if it were, p would not be justified for S on the basis of w, contrary to the hypothesis that the antecedent in the Contrary Consequence Elimination Principle is fulfilled. If ∼ p were a grounded or pseudogrounded overriding proposition, then there would be an overriding proposition for the evidence, w_1, for p, since $(w_1 \& \sim p) \mathcal{C} p$. But if the antecedent of the Contrary Consequence Elimination Principle is fulfilled, S must be justified in believing that p. In other words, if Jsp, and p → q, there can be no overriding proposition of the evidence, p, for q. Thus, the third, and last, of the conditions sufficient for q to be justified for S obtains — there is no overriding proposition, u, of q for S.

We have shown that if p is justified for S, and p entails q, then q is justified for S. For we have shown that p is grounded and confirms q, and we have shown that there can be no overriding evidence of p for q. And according to our model of justification, that is sufficient to guarantee that q is justified for S.

Thus there appears to be a perfectly general argument for the Principle of the Transmissibility of Justification Through Entailment, and I believe that there is only one plausible objection which, when examined carefully, presents no real difficulty. It could be argued that my reason for the claim that there is no grounded overriding proposition of the evidence, p, for q depends upon rejecting a claim I made earlier. For I had claimed that if p → q, then pCq. Now, I also said that the only possible overrider of the evidence, p, for q which we need to consider would be ∼ p. That is, I claimed that $(p \& \sim p) \mathcal{C} q$. On the other hand, since $(p \& \sim p) \rightarrow q$, the proposition ∼ p cannot override the evidence, p, for q, if I continue to maintain that for all x and y, if x → y, then xCy.

This objection strikes me as well founded; and the restriction mentioned earlier needs to be imposed upon the general claim that if x → y, then xCy. For not only do contradictory propositions raise difficulties with the general unrestricted claim that if x entails y, then x is adequate evidence for y, but necessarily true propositions raise equally serious problems. That is, it hardly seems that contradictory propositions provide adequate confirming evidence for every proposition; and it seems equally implausible that every proposition provides adequate confirming evidence for a necessarily true proposition. Yet both of those consequences follow from the unrestricted claim. Thus, some may argue that *if* this principle is to apply to necessarily false and necessarily true propositions, some restrictions need to be incorporated; and I agree.

An appropriate restriction to eliminate contradictory propositions as adequate confirming evidence is not as easily formulated as it may at first appear. We could add a conjunct to the antecedent of the general claim; for example, if x entails y, and there is some z such that x does not entail z, then xCy. Thus, our revised principle would not count contradictory propositions as adequate confirming evidence. But that would, I fear, rule out by definition the possibility of *any* proposition being justified for many S's. Suppose that S_1 believes that r and believes that $\sim r$. Presumably, many of us have contradictory beliefs. Now, since both r and $\sim r$ are in B_{S_1}, and since we allowed propositions to be "put together" to form overriders, the proposition $(r\ \&\ \sim r)$ could be constructed from S's beliefs. But if our general rule were amended to rule out all contradictory evidence as confirming evidence, the proposition $(r\ \&\ \sim r)$ would override every confirmation. For example, if e is in $\Gamma^R_{S_1}$ and there is a nondegenerate chain from e to p, p would be grounded but not justified, since $[e\ \&\ (r\ \&\ \sim r)]\ \mathcal{C}p$. Thus, the general claim that 'if $x \rightarrow y$, then xCy' cannot be amended by simply disallowing contradictory propositions as adequate confirming evidence.

The other problem is equally difficult to solve, since it is not immediately clear what we should count as adequate confirming evidence for a necessary truth. Can contingent propositions serve as adequate confirming evidence for a necessary truth? Suppose an expert sincerely testifies that a certain proposition is a theorem. Does that provide evidence for the theorem? Or must the evidence itself be a necessary truth?

Fortunately, these are problems which we can leave unsolved, because the arguments for scepticism which we have chosen to examine concern the possibility of knowledge of contingent empirical propositions. Thus, our rules of confirmation need not be sufficiently general to encompass necessarily true propositions or necessarily false propositions. Specifically, the issue here is whether S is justified in believing that q, if p entails q and S is justified in believing that p. Neither p nor q are taken to be necessarily true or necessarily false propositions. We can specifically exclude necessarily true and necessarily false propositions from the domain of our variables which range over propositions.

Thus, we can take the variables in the claim that 'if x → y, then xCy' to range over only contingent propositions and nevertheless hold that $(x \& \sim x)\mathcal{C}y$ and $(x \& \sim y)\mathcal{C}y$. The restrictions are acceptable simply because $\sim x$ is an overrider of x for any value of x, including, of course, contingent propositional values, and $(x \& \sim y)$ fails to confirm y no matter what the values of x or y, including contingent values. Our model of justification need only be sufficiently general to allow us to test the Transmissibility Principle when all the propositions are contingent. By restricting the range of propositions to contingent propositions, we can maintain both that 'x → y, then xCy' and that '$(x \& \sim x)\mathcal{C}y$'.

I take it, then, that neither contradictory propositions nor necessarily true ones raise any real difficulty here. But it may be thought that the restriction introduced to block contradictions from providing adequate evidence is ad hoc, i.e., introduced only to make my account consistent. It is crucial to note, however, that it is not even *necessary* to adopt any restriction! For whether or not this restriction is adopted, the purported objection to my derivation of the transmissibility principle from the Model of Justification will fail. It will not show that there is any grounded proposition which overrides the evidence, p, for q.

Whatever restriction we adopt will allow $\sim p$ to override the evidence, p, for q or it will not allow it. If it permits $\sim p$ to override —that is, if $(p \& \sim p)\mathcal{C}q$—then q will not be justified for S, but, as I have argued, neither will p be justified. For $\sim p$ also overrides the evidence, w, for p. Now, since $\sim p$ is the only possible overrider of the evidence, p, for q which we need to consider (because p → q), if the consequent of the transmissibility principle is not fulfilled because there is an overrider, neither is the antecedent because of the same overrider.

On the other hand, if the restriction which we adopt does not allow $\sim p$ to override—that is, if $(p \& \sim p)Cq$—the third necessary condition of q being justified for S would, once again, be fulfilled. For (since $\sim p$ is the only possible overrider of the justification of q which we need to consider) there would be no overriding proposition of the evidence, p, for q.

I believe that we can conclude that there is a good, apparently unobjectionable argument for the Principle of the Transmissibility of Justification Through Entailment. And as mentioned above, if that principle is correct, the Contrary Consequence Elimination Principle must also be correct, since it is merely a corollary of the Principle of the Transmissibility of Justification through Entailment. In order to avoid the appearance of cooking up a general model of justification which has just the consequences needed for my defense of the Principle of Transmissibility, however, it would be useful to show that it has several other desirable features. The argument for each of these is similar to the argument for the transmissibility principle, and I will only sketch them here.

The Conjunction Principle follows from the model; that is, 'If S is justified in believing that (x&y), then S is justified in believing that x, and S is justified in believing that y' is merely a corollary of the transmissibility principle. For (x&y) entails x and it entails y. The converse of that principle does not hold, because there can be overriders of the proposition (x&y) which are not overriders of either x or overriders of y. This is surely a welcome result, for it points to a partial solution of the Lottery Paradox. The paradox can be put as follows: Suppose that S believes that a certain lottery has one hundred tickets and that only one ticket will win. Many people would argue that S is justified in believing with regard to each ticket that it will not win. That is, S is justified in believing that t_1 (ticket no. 1) will not win, and S is justified in believing that t_2 will not win, and S is justified in believing that t_3 will not win, and . . . S is justified in believing that t_{100} will not win. But they would also argue that S is not justified in believing the conjunction '(t_1 will not win & t_2 will not win & t_3 will not win & . . . t_{100} will not win)' because S knows that at least one of the tickets will win. The proposed model of justification can account for those intuitions, since there is an overrider of the conjunction '(t_1 will not win & t_2 will not win & . . . t_{100} will not win') which is not an overrider of any conjunct, namely '~ (t_1 will not win & t_2 will not win & . . . t_{100} will not win').[25] We will consider the Lottery Paradox more fully in section 3.15.

The Disjunction Principle follows from the Model of Justification; that is, 'If S is justified in believing that x or S is justified in believing that y, then S is justified in believing that (xvy)' is also a corollary of the Transmissibility Principle. The converse is not true, since there may be confirming evidence of (xvy) which is not confirming evidence of x or confirming evidence of y. For example, there may be a grounded proposition sufficient to confirm that either Smith or Wesson is the murderer without it being the case that those grounded propositions either confirm that Smith is the murderer or confirm that Wesson is the murderer.

The Contradictory Exclusion Principle follows from the Model of Justification; that is, 'If S is justified in believing that x, then S is not justified in believing that ∼ x' can be demonstrated by a simple reduction argument. Suppose that S is justified in believing that x and justified in believing that ∼ x. I have already shown that if a proposition is justified for S, it is grounded for S. Hence, ∼ x is grounded for S. But if ∼ x is grounded, there is an overriding proposition of the confirming evidence, whatever it is, for x, namely, ∼ x. Hence, contrary to our assumption, x is not justified for S. Thus, if S is justified in believing that x, S is not justified in believing that ∼ x. Note that nothing said here prevents S from having adequate grounded confirming evidence for x and adequate grounded confirming evidence for ∼ x.

Finally, the Principle of the Nontransitivity of Justification can be shown to follow from the Model of Justification. That principle asserts that if x justifies y for S, and y justifies z for S, it does not follow that x justifies z for S. In our terminology that principle means that (1) if S is justified in believing that y because xCy and there is no overrider of x for y *and* (2) if S is justified in believing that z because yCz and there is no overrider of y for z, then it does not follow that S is justified in believing that z because xCz and there is no overrider of x for z. The antecedent of the principle guarantees that there is a nondegenerate chain terminating in '. . . xCy' and one terminating in '. . . yCz', but it does not guarantee that there is a nondegenerate chain terminating in '. . . xCz'. However, since S is justified in believing that z because yCz and there is no overrider of y for z, then there must be a nondegenerate chain terminating in '. . . xCyCz'. That is so because if there are *no* overriders of y for z, there can be no *internal* overriders of y for z. But what follows from this is not that xCz, but rather that xC*z; that is, x is in the confirming ancestry of z and provides a confirmation of z because it confirms another proposition, y. If the only nondegenerate chain available to S terminating in z is '. . . xCyCz', then we can say that S is justified in believing that z *only because* xCy; for if x₵y, S would have no pure chain terminating in z. The notion of a proposition's being justified (or not justified) for S *only because* another proposition is confirmed is important, and I will return to it in Chapter Three when considering the distinction between misleading and genuine defeaters. That distinction, in turn, is crucial to the account of knowledge and certainty to be developed in Chapter Three.

The results of this section can be summarized as follows: Several principles of justification follow from the Model of Justification developed in section 2.7. The most important one for our purposes is the Principle of Transmissibility of Justification Through Entailment, because it is one of the interpretations of the sceptic's Basic Epistemic Maxim employed in the Evil Genius Argument. In addition, since other desirable principles of justification follow from the same features of the Model of Justification which lead to the transmissibility principle, that model and the transmissibility principle, itself, seem substantiated.

2.9 Summary of the Analysis of the Evil Genius Argument up to This Point

Let me summarize the results of our analysis of the Evil Genius Argument up to this point. We saw that one interpretation of the sceptic's Basic Epistemic Maxim that 'in order for S to be justified in believing that p, S must be justified in denying the Evil Genius Hypothesis' was: If S is justified in believing that p (where 'p' stands for some proposition which we believe is knowable), then S is justified in denying that H_c (where 'H_c' is a logical contrary of p). We took this to be an instantiation of the more general epistemic principle, the Contrary Consequence Elimination Principle, namely, if x and y are logical contraries, and S is justified in believing that x, then S is justified in believing that $\sim y$. That principle, in turn, is a corollary of the even more general epistemic principle, namely if x entails y, and S is justified in believing that x, then S is justified in believing that y. We called that the Principle of the Transmissibility of Justification Through Entailment.

In the previous sections, I argued that the objections to the transmissibility principle were inconclusive. In addition, I argued for the principle, restricted in appropriate ways, by using what I take to be an acceptable general model of justification which has the transmissibility principle as one among many welcomed consequences. Along the way I developed an account of the propositions which are available for S to use as confirming evidence and an account of propositions available to S which override those confirming propositions. Both of those accounts will be useful during our discussion of certainty, which is postponed until Chapter Three. In addition, the Model of Justification developed in these sections will be employed to help us evaluate other interpretations of the sceptic's Basic Epistemic Maxim.

But the task, now, is to see how the sceptic makes use of the Contrary Consequence Elimination Principle. Specifically, can the sceptic employ it to motivate the conclusion that we are not justified in believing those propositions which we normally believe are knowable?

2.10 The Contrary Consequence Elimination Principle, Though True, Does Not Provide a Reason for Direct Scepticism

We investigated the Contrary Consequence Elimination Principle in order to show, among other things, that it was correct, since the sceptic appears to rely heavily upon it in formulating the argument for Direct Scepticism—and our strategy is to give the sceptic the best run for his/her money. But the crucial question to ask now is: Given that the principle is correct, does it really lend support to the sceptic's position? I think that it does not. In fact, I believe that it is equally useful (if not more so) to the nonsceptic!

The argument below, using the Contrary Consequence Elimination Principle, is probably what the sceptic has in mind. (Often the second premiss is unstated.)

S1.1 If S is justified in believing that p (where 'p' stands for a proposition believed to be knowable), then S is justified in believing that there is no epistemically malevolent mechanism making him/her (S) falsely believe that p (instantiation of the Contrary Consequence Elimination Principle).

S1.2 S is not justified in believing that there is no epistemically malevolent mechanism making him/her (S) falsely believe that p.

∴   S is not justified in believing that p (by modus tollens).

The argument can be represented schematically as follows:

S1.1 $Jsp \rightarrow Js \sim H_c$

S1.2 $\sim Js \sim H_c$

∴ $\sim Jsp$

What should we say about this argument? It certainly is valid (i.e., the conclusion follows necessarily from the premisses), and, as we have seen, S1.1, or something very much like it, is true. But what of S1.2? Has any evidence been given to support it? And what would that evidence be?

Consider the counterargument:

S1.1 $Jsp \rightarrow Js \sim H_c$ (same as in the sceptical argument)

C1.2 Jsp (what the nonsceptic believes to be true)

∴ $Js \sim H_c$ (by modus ponens)

Now, is S1.2 true? Or is C2.2 true? They cannot both be true (because of the Contrary Consequence Elimination Principle). What grounds could the sceptic have for asserting that S is not justified in rejecting H_c? Granted it is logically possible that H_c. But that cannot be a reason for asserting S1.2, since the sceptic surely cannot require that only logically necessary propositions are justified. It cannot be claimed that it is epistemically possible that H_c and, therefore, S is not justified in rejecting H_c, since that would clearly beg the question. For to show that H_c is compatible with everything known by S, it would first have to be determined whether S knows that p, since H_c is incompatible with p.

Thus, although S1.1 (or the Contrary Consequence Elimination Principle) is usually the focus of attention, I believe that the crucial step in this argument is S1.2. Surely the sceptic cannot simply claim that it is *obvious* that S is not justified in rejecting H_c. But, what is more important, I believe that the sceptic will be forced to argue that S is not justified in believing that p in order to argue for the claim that S is not justified in rejecting H_c.

In order to see this last, crucial point, consider what the argument between the sceptic and nonsceptic must look like. The nonsceptic would certainly claim that it is at least equally obvious that S is justified in believing that p as it is that S is not justified in rejecting H_c. In addition, the nonsceptic would continue, since p is adequate confirming evidence for rejection H_c, S *is* justified in rejecting H_c. We saw in the previous sections that, in supporting S1.1 against the Dretske-like attacks, the sceptic would be forced to grant that the adequate confirming evidence, e, for p need not be adequate confirming evidence for $\sim H_c$. Specifically, in order to save the Contrary

Consequence Elimination Principle from attacks, it was necessary to argue that confirmation is not transitive. In the Zebra Case, *the* evidence which confirmed the proposition that the animals are zebras did not confirm the proposition that the animals are not painted mules. The proposition which confirmed that they are not painted mules was the proposition that they are zebras. Consequently, in the Evil Genius Case, the fact that e is not adequate confirming evidence for $\sim H_C$ could not show that S is not justified in denying H_C.

It can be granted to the sceptic that e is not adequate confirming evidence for denying H_C. But there *is* adequate evidence which *may* be available for S for $\sim H_C$, namely, p itself. In addition, as the counterargument shows, the nonsceptic will claim that *the* sufficient condition for being justified in denying that H_C which does obtain is that S is justified in believing that p. Consequently, if the sceptic has an argument for denying that S is justified in believing that $\sim H_C$, that argument must show that p is either not confirmed by evidence available to S or that, if it is, it is overridden. But that is just the issue at stake! For it is equivalent to the *conclusion* of the sceptic's argument using the Contrary Consequence Elimination Principle. That is, it is equivalent to the proposition that S is not justified in believing that p. It certainly cannot be used as evidence for the claim that S is not justified in believing that $\sim H_C$.

In order to clarify my claim that this argument for Direct Scepticism begs the question, it will be useful to return to the model of justification developed in section 2.7 and the argument for the Contrary Consequence Elimination Principle presented in section 2.8. That model of justification can be summarized as follows:

S is justified in believing that $x =_{df}$

1. There is a nondegenerate chain of grounded propositions anchored in Γ_S^R and terminating in a proposition, w, such that wCx;

and

2. There is no grounded or pseudo-grounded proposition, u, for S such that u overrides the confirmed evidence, w, for x.

In order to show that S is not justified in believing a proposition, say x, it would have to be shown that one of the necessary conditions of justification fails to obtain. There are three possible strategies available:

Strategy I   There is no pure chain of grounded propositions anchored in Γ_S^R terminating in a proposition, w, such that wCx;

or

Strategy II   If there is such a chain, there is an available overrider, u, of the confirmed proposition x;

or

Strategy III There is a proposition, u, such that for every S it would override any adequate confirming evidence for a proposition, x, located at the end of a pure chain anchored in Γ_S^R.

In establishing the Contrary Consequence Elimination Principle, it was shown that if x entails y, and S is justified in believing that x, then there is a nondegenerate chain of grounded propositions anchored in Γ_S^R terminating in x which confirmed y and that there was no overrider for S of the confirmed proposition y. In addition, I argued that if there were any overrider of x for y, then $\sim$ x would be available for S as an overrider. That was so because, since x entails y, any overrider of x for y (i.e., either $\sim$ x or $\sim$ y) would make $\sim$ x available as an overrider to S.

Let me apply these very general results to the particular case before us. To repeat: in order for the argument which employs an instantiation of the Contrary Consequence Elimination Principle to provide a reason for Direct Scepticism, the second premise needs to be substantiated. That is, the sceptic must provide an argument for the claim that S is not justified in believing that $\sim H_c$. In order to avoid begging the question, that argument must not depend upon showing that S is not justified in believing that p, since that is the conclusion of the argument for Direct Scepticism employing the Contrary Consequence Elimination Principle. But the success of each of the three strategies for showing that S is not justified in believing that $\sim H_c$ would depend upon showing that S is not justified in believing that p.

Strategy I If the sceptic is going to show that there is no pure chain of grounded propositions anchored in Γ_S^R and terminating in a proposition which confirms $\sim H_C$, the sceptic will have to show that S is not justified in believing that p. For if p were justified for S, there would be such a chain—namely, the one terminating in p.

Strategy II This option is no better for the sceptic. Since a pure chain of grounded propositions terminating in p and anchored in Γ_S^R would confirm $\sim H_C$, and since p entails $\sim H_C$, $\sim$ p would have to be available to S as an overrider if there is to be an overrider of p for $\sim H_C$. Now, of course, $\sim$ p may be available to some S's since it may be actually subscribed to by some S's. In such a case, it would be in B_S for those S's and, hence, available as an overrider. It may also be a conjunction of propositions each of which is either in B_S or a link of a pure chain anchored by a conjunction of propositions each of which is in Γ_S or Γ_S^R or A_S. But the sceptic must show that $\sim$ p is available as an overrider in one of the above ways for every S who has a pure chain terminating in p and anchored in Γ_S^R. For the sceptic's claim that S is not justified in believing that $\sim H_C$ must apply to every S—if the argument employing the Contrary Consequence Elimination Principle is to provide a reason for accepting Direct Scepticism. The minimal requirement that must be fulfilled if $\sim$ p is available to every S who has a pure chain anchored in Γ_S^R terminating in p is that no S is justified in believing that p. (S could not be justified in believing that $\sim$ p, because of the pure chain ending in p.) Thus, the sceptic must once again show that S is not justified in believing that p.

Strategy III Similarly, adopting this final strategy will force the sceptic to beg the question. For if there is an overriding proposition, u, available to *every* S such that u would override any adequate confirming evidence for $\sim H_C$ at the end of a pure chain anchored in Γ_S^R, u would have to be a conjunction at least one conjunct of which is $\sim$ p. Again, the minimal condition assuring that there is such an overrider for every S is that no S is justified in believing that p. And, once again, the sceptic would have to show that S is not justified in believing that p.

To sum up: the Contrary Consequence Elimination Principle will not help the Direct Sceptic. For it is just as useful to the nonsceptic in asserting that S is justified in believing that there is no evil genius making it only appear that p. And, if the sceptic is going to argue that S is not justified in rejecting H_C, he/she will have to show that

S is not justified in believing that p. But that is what the argument using the Contrary Consequence Elimination Principle was supposed to show.

Now, that principle is only one of four interpretations of the sceptic's Basic Epistemic Maxim. It portrayed the basic maxim as asserting a logically necessary relationship between being justified in believing that p and being justified in rejecting H_c. It does not require that S be justified in rejecting H_c *before* S is justified in accepting p. I will return to that suggestion later (see sections 2.13 through 2.16).

2.11 Irrationality and Dogmatism: Two Misinterpretations of the Contrary Consequence Elimination Principle

Aside from the argument examined in the preceding section, there are two ways in which the Contrary Consequence Elimination Principle may be thought to lead to Direct Scepticism. I believe that both involve serious misunderstandings of that principle. The first conflates the distinction between the *Contrary* Consequence and the *Defeater* Consequence Elimination Principles. The second involves a misunderstanding of the entailment between the antecedent and consequence in the Contrary Consequence Elimination Principle.

The first way in which it may be thought that the Contrary Consequence Elimination Principle leads to Direct Scepticism is that it may appear that it is "epistemically foolhardy" or "irrational" for S to deny that the evil genius is up to his/her old tricks. And if it is *irrational* to believe that the evil genius is not making S falsely believe that p (where 'p' is some proposition ordinarily believed to be knowable, e.g., 'there are rocks'), then S cannot be justified in believing that the evil genius is not making him/her (S) falsely believe that p. And by the Contrary Consequence Elimination Principle, if S is not justified in denying H_c, then S is not justified in believing that p. The sceptic may claim that the proposition that there are rocks could not possibly confirm the proposition that there is no evil genius making S falsely believe that p. In the last three sections I argued that it did.

Consider what Peter Unger says:

. . . [An] attempt to reverse our argument [the sceptic's Evil Genius Argument] will proceed like this: According to your argument, nobody ever *knows* that there are rocks. But, I *do* know that there are rocks. This is something concerning the external world, and I do know it. Hence, somebody *does know* something about the external world. Mindful of our first premiss [a premiss similar to what I have given as an instance of the Contrary Consequence Elimination Principle], the reversal continues: I can reason at least moderately well and thereby come to know things which I see to be entailed by things I already know. Before reflecting on classical arguments such as this, I may have never realized or even had the idea that from there being rocks it follows that there is *no* evil scientist [Unger's evil genius] who is deceiving me into *falsely* believing there to be rocks. But, having been presented with such arguments, I of course *now* know that this last *follows* from what I know. And so, while I might not have known *before* that there is no such scientist, at least I *now* do know that there is no evil scientist who is deceiving me into falsely believing that there are rocks. So far has the sceptical argument failed to challenge my knowledge successfully that it seems actually to have occasioned an increase in what I know about things.

While the robust character of this reply has a definite appeal, it also seems quite daring. Indeed, the more one thinks on it, the more it seems to be somewhat foolhardy and even dogmatic. One cannot help but think that for all this philosopher really can *know*, he might have all his experience artificially induced by electrodes, these being operated by a terribly evil scientist who, having an idea of what his 'protégé' is saying to himself, chuckles accordingly. One thinks as well that for all *one can know oneself,* there really is . . . no other thinker with whose works one has actually had any contact. One's belief that one has may, for all one really can *know,* be due to experiences induced by just such a chuckling operator. For all one can *know,* then, there may not really be any rocks. Positive assertions to the contrary, even on one's own part, seem quite out of place and even dogmatic.[26]

Unger summarizes his objection to S's being justified in denying the Evil Genius Hypothesis as follows:

No matter what turns one's experience takes, the statement that one *knows* there to be no scientist *may* be wrong *for the reason that there is a scientist* [Emphasis added.] But, it *will always* be wrong, it seems, for the reason of dogmatism and irrationality, however this last is to be explained.[27]

We will consider the charge of dogmatism shortly. The question before us now is whether the belief that there is no evil genius making

S falsely believe that there are rocks is foolhardy and irrational.

Although Unger is here speaking of "knowledge" rather than one of its necessary conditions, justification, presumably the purported "foolhardy" and "irrational" character of the nonsceptic's position depends upon the nonsceptic maintaining that he/she has a justification of the claim that there is no evil genius making him/her falsely believe that there are rocks. Now, if the nonsceptic employed the proposition that there are rocks to confirm such propositions as *there is no evil scientist* or *there is no evil scientist who could make S believe a false proposition*, as Unger implies in his summary, he/she would, indeed, be embarking on an epistemically foolhardy course. But it is crucial to note here, as we did when we considered the Appointment Case (see section 2.5), that these propositions are not entailed by the proposition that there are rocks. For it is certainly possible for there to be rocks and for there to be an evil scientist with the power to make S believe falsely that there are rocks, if there were no rocks. The proposition that there is such a scientist with such powers may be an overrider of the evidence for the claim that there are rocks, but it is not a logical contrary.[28]

To see this, reconsider the sceptical hypothesis embodied in H_c. It includes the proposition that there is no evil scientist who is making S falsely believe that p. That proposition includes the conjunct that S falsely believes that p. Finally, that is merely the conjunction of 'S believes that p' and 'p is false'. Thus, the sceptical hypothesis includes the conjunct '$\sim p$'. In fact, we have it expressed it this way:

H_c e (the evidence for p) & $\sim p$ & there is some malevolent mechanism, M, which brings it about that S believes (falsely) that p.

I pointed out that '$\sim p$' was listed separately (although it was entailed by the third conjunct) in order to emphasize the difference between H_c and another sceptical hypothesis, H_d. That hypothesis is:

H_d e & there is some mechanism, M, which could bring it about that S falsely believes that p (without changing the truth of 'e')

H_c, but not H_d, is a contrary of p, precisely because H_c, but not H_d, contains '$\sim p$' as a conjunct. Since the Contrary Consequence Elimination Principle is correct, p would confirm the denial of *any* conjunction one conjunct of which is $\sim$ p. Thus, the inference to $\sim H_c$ by S is neither foolhardy nor irrational. The inference is an absolutely trivial one. If S is justified in believing that there are rocks, then S is justified in believing that there is no evil scientist making S *falsely* believe that there are rocks, not because S is justified in believing that there is no scientist with extraordinary powers, but rather because S is justified in denying *any* conjunction containing $\sim$ p. S has not miraculously "increased" his/her knowledge, as Unger would suggest. Quite the contrary: S would be absolutely justified in believing that $\sim (\sim p \& x)$ no matter what proposition were put in the place of x. Thus, S is justified in believing that there is no evil scientist making him/her falsely believe that p simply because S is justified in believing that $\sim$ p is false. S does not have to have *any* evidence about an evil scientist or about the powers of such a scientist. But S is justified in denying that H_c, if S is justified in believing that p.

The Contrary Consequence Elimination Principle merely sanctions the rejection of every logical contrary of p. The Defeater Consequence Elimination Principle, on the other hand, would sanction the rejection of the proposition that there is such an evil scientist. But as we will see, it is because the Defeater Consequence Elimination Principle *would* sanction such a foolhardy inference that it should be rejected. We will consider that principle shortly.

Let us turn to the second, related way in which the Contrary Consequence Elimination Principle (or, for that matter, the Defeater Consequence Elimination Principle) might be thought to lead to Direct Scepticism. It is that it may seem to sanction the right to be dogmatic. It might be held that since 'S knows that p' entails 'S is (completely) justified in believing that p', (and since the latter entails 'S is justified in rejecting *any* incompatible proposition, q' knowledge leads to the epistemic right to close one's mind to any future incompatible evidence, i.e., the right to be dogmatic.[29] But since one never has the epistemic right to be dogmatic, the objection would continue, S cannot know that p.

Although a full analysis of this would require us to examine the nature of certainty, which is postponed until Chapter Three, the point I wish to make here is that this defense of scepticism relies upon a serious misunderstanding of the elimination principles. Although, as we have just seen, there are important differences between the two elimination principles, they are similar in that neither leads to dogmatism. I will explicitly defend only the Contrary Consequence Elimination Principle against that charge; but the argument could easily be converted to defend the other Elimination Principle.

The Contrary Consequence Elimination Principle (which I have already shown to follow from an acceptable partial characterization of justification) involves the assertion that, necessarily, if S is justified in accepting p, S is justified in rejecting any incompatible (i.e., contrary) proposition, q. Thus, S has the epistemic right to deny any proposition incompatible with p, *if and whenever* S knows that p (since being completely justified in believing that p can be granted as a necessary condition of knowledge).[30] But *if and whenever* S becomes aware of some potentially incompatible evidence, q, S may lose *both* the justification for p and the right to reject q. Nothing prevents such conflicting evidence coming to S's attention; in fact, nothing prevents S from seeking such evidence.

To make this point more clear, consider a parallel principle. Call it the *Person Location Principle* (PLP):

Person Location Principle (PLP)	For any two places P_1 and P_2 such that $P_1 \neq P_2$, necessarily if S is located in P_1, S is located in non-P_2 (where 'non-P_2' indicates the set of "contrary" places to P_2).

Now, PLP does not require that S, having once located himself/ herself at P_1, remain forever glued (so to speak) to P_1. Nothing prevents S from seeking to leave P_1 or seeking while still at P_1 to go to P_2. Similarly, the Contrary Consequence Elimination Principle does not prevent S from having a mind open to counterevidence, nor does it prevent S from seeking that evidence. Even if S knows that he/she is (completely) justified in believing that p and knows that he/she has the right to reject counterevidence *as long as he/she remains justified in believing that p,* it would not be impossible, irrational, or epistemically perverse for S to seek counterevidence for p. S may seek to discover more about the state of affairs represented by 'q' and, at the instant S thinks that q is plausible, S is no longer completely justified in believing that p. Just as S can move from P_1 to P_2, the epistemic status of the belief that p (and the corresponding rights) may change as a result of acquiring new evidence incompatible with p.

2.12 The Defeater Consequence Elimination Principle is False, but, Even if It were True, It Would Be Equally Useless to the Sceptic

The conclusion of the two preceding sections was that the Contrary Consequence Elimination Principle, although true, does not provide any warrant for Direct Scepticism without some additional evidence for the claim that S is not justified in rejecting H_c. The issue is not the truth of that general epistemic principle but rather the evidence for the claim that S is not justified in denying H_c. Without some independent supporting argument for that claim, the sceptic will have failed to provide any reason for Direct Scepticism. But that "independent" argument was shown to beg the very issue in question.

Let us now turn our attention to the other elimination principle—the Defeater Consequence Elimination Principle. Here the claim is that for all propositions (x, y) if y is a defeater of S's justification of x, then, if S is justified in believing x, then S is justified in rejecting y. The Defeater Consequence Elimination Principle is a more general principle than the elimination principle considered earlier, because all incompatible propositions are defeaters, but not all defeaters are incompatible propositions. The argument making use of it is analogously more general.

S2.1　If S is justified in believing that p (where 'p' stands for a proposition believed to be knowable), then S is justified in believing that there is no epistemically malevolent mechanism which could bring it about that S believes falsely that p (instantiation of the Defeater Consequence Elimination Principle).

S2.2　S is not justified in believing that there is no epistemically malevolent mechanism which could bring it about that S believes falsely that p.

∴　　S is not justified in believing that p (by modus tollens).

The argument can be represented schematically as follows:

S2.1　$Jsp \rightarrow Js \sim H_d$

S2.2　$\sim Js \sim H_d$

∴　　$\sim Jsp$

First, many of the previous considerations with regard to the Contrary Consequence Elimination Principle apply here as well. That is, the Defeater Consequence Elimination Principle, even if true, would provide no grounds for accepting the conclusion that S is not justified in believing that p unless some reason can be given for S2.2. In addition, for the reasons already considered, the Defeater Consequence Elimination Principle does not sanction dogmatism. But there are unavoidable difficulties with this elimination principle. For, whereas the Contrary Consequence Elimination Principle can be supported by a general argument, this more general elimination principle does not appear correctly to characterize justification, and as we just saw (in section 2.11), it would sanction epistemic recklessness.

This principle asserts that if S is justified in believing that x, S is justified in rejecting all defeaters of the justification of x.[31] We have already considered sufficient reasons for rejecting this principle. Recall the Appointment Case and Unger's Rocks-and-Evil-Scientist Case. In both of them, it was clear that there were defeaters of justified beliefs for S which S was not justified in denying. Specifically, recall that even though S may be justified in believing that Ms. Reliable will keep the 2:00 P.M. appointment, S may not be justified in believing that Ms. Reliable will not receive a phone call prior to 2:00 P.M. in which she is told that her apartment building has burned down. Also, even though S may be justified in believing that there are rocks, S may not be justified in believing that there is no evil scientist who could make him/her (S) believe that there are rocks even if there were no rocks.

Elsewhere I have used one other standard case which appears in the literature to show that the Defeater Consequence Elimination Principle is false.[32] Consider the famous Tom Grabit Case.[33] Recall that S knows Tom fairly well and that S observes what he/she takes to be Tom stealing a book from the library by cleverly concealing it underneath his coat. Lacking any overriding counterevidence, it appears that, according to the Model of Justification developed in section 2.7, S is justified in believing that (t) Tom stole the book.

Now, consider the following: it seems that many persons have at least one double. That is, there is probably at least one person who looks sufficiently like Tom, at a casual inspection, to be indistinguishable from Tom at the distance from which S observed Tom. Call Tom's double "Dom." 'Dom was in the library on the day in question', if true, is surely a genuine defeater of S's justification for t. Now, if the Defeater Consequence Elimination Principle were correct, and if S were justified in believing that t, S would be justified in believing that Dom was not in the library on the day Tom stole the book. But what justification does S have for that? Does S have any idea at all about Dom's whereabouts? S may not even know who Dom is or that Tom has a double. If S were to conclude that Dom was not in the library on the basis of his/her belief that Tom was in the library, S would surely be epistemically foolhardy. Of course, it is not likely that a double-for-casual-inspection-purposes would be in the vicinity of Tom; but presumably that is because the probability of Dom being in *any* particular place is very small. Tom's whereabouts give us no evidence about Dom's whereabouts. In other words, 'Tom was in the library' does not confirm 'Dom was not in the library', and, hence, S does not have a justification for the latter.

As a further test of the Defeater Consequence Elimination Principle, consider whether S *knows* that Dom was not in the library. For if the proposition that Dom was not in the library was true, justified, and believed (and S's justification is nondefective), S would *know* that Dom was not in the library. To grant S a justification would be to grant knowledge, other things being equal. Even the opponent of the sceptic would not wish to enlarge the corpus of knowledge by that much! Thus, the Defeater Consequence Elimination Principle does not appear to characterize justification correctly.

The sceptic, however, may reply that the case involving Tom and Dom is sufficiently similar to the Evil Genius Case so that if we are not justified in rejecting the claim that Dom was in the library, we are not justified in claiming that Tom stole the book. That is, the sceptic might claim that the example does not show that the Defeater Consequence Elimination Principle is incorrect, but rather that the presupposition is false that we are justified in believing that t.

What are we to say in reply? First, it is important to note that the considerations which protected the Contrary Consequence Elimination Principle against the Dretske-like attacks will not aid in the defense of the Defeater Consequence Elimination Principle. Reconsider the Zebra Case (in section 2.5). Suppose that everything is as it was except that the zoo keeper has frequently substituted painted mules for the zebras. On *this* particular occasion, he is not up to his old tricks: the zebra-looking animals are zebras. Further, suppose that S knows nothing about the previous exchanges of zebra facsimiles for zebras. Thus, 'the zoo keeper has frequently substituted painted mules for zebras' is a defeater of S's justification for the claim that there are zebras. As in the original case, the *evidence* which S has for the claim that there are zebras is not sufficient for denying the defeater. *And* since the claim 'there are zebras in the zoo (now)' is not *itself* adequate (or even close to being adequate) confirming evidence for denying the defeater, the argument against the Contrary Consequence Elimination Principle, although unsuccessful there, can be used against the Defeater Consequence Elimination Principle. The reason is that although a proposition is adequate confirming evidence for the denials of all of its contraries, it is not adequate confirming evidence against all defeaters of its justification.

Thus, the Dretske-like considerations seem persuasive against the Defeater Consequence Elimination Principle. But, as I said above, the Tom-Dom Case and now the revised Zebra Case so closely resemble the Evil Genius Case that I suspect that the sceptic may not be persuaded to abandon this principle on their account. Nevertheless, they do point to a general dilemma for the sceptic. For after all, there do *seem* to be some cases similar to the Evil Genius Case in which S is justified in believing that p but not justified in denying some of the defeaters of that justification (i.e., the cases just considered). The problem for the sceptic is that since this principle, unlike the other elimination principle, is itself suspect, any evidence *for* the claim that S is not justified in rejecting H_d (i.e., S2.2) would seem to count *equally* against this principle. Thus, the sceptic will be unable to establish both of the premisses of the argument using the Defeater Consequence Elimination Principle. Evidence *for* the second premiss casts further doubt *against* the first premiss.

However, let us suppose that this principle is correct. As we saw in examining Dretske's argument against the Contrary Consequence Elimination Principle, this supposition would require adopting the assumption that p is adequate confirming evidence for denying any defeater of the justification for p. Specifically, we would have to assume that p is adequate confirming evidence for denying H_d. It is that assumption which surely seems false. But let us grant it. In that case, this principle, like the other elimination principle, would become equally valuable to the sceptic *and* his/her opponent. For the opponent will claim that there is a counterargument against S2.2 just as there is against S1.2. The considerations raised earlier to show that the Contrary Consequence Elimination Principle is not more useful to the sceptic than to his/her opponent will be applicable here as well. Those considerations will probably not appear to be as convincing, because the Defeater Consequence Elimination Principle is false. But if that principle were true, it would not be foolhardy or irrational to conclude, in the Appointment Case, for example, that Ms. Reliable will not receive the distressing phone call. And it would not be unreasonable for S to conclude, in the Rocks-and-Evil-Scientist Case, that there is no evil scientist. Similarly, it would not be unreasonable for S to conclude in the Tom-Dom Case, that Dom was not in the library. If the Defeater Consequence Elimination Principle were correct, the inferences to the denials of the defeaters would be justified. Thus, returning to the argument for Direct Scepticism employing that principle, the sceptic would have to show that S is not justified in believing that p in order to show that S is not justified in denying H_d. That is, if the Defeater Consequence Elimination Principle were correct, the sceptic will have to show that S is not justified in believing that p in order to substantiate the claim in S2.2. But, once again, that argument will beg the very issue at stake, i.e., whether S is justified in believing that p.

In summary, we can conclude that the Defeater Consequence Elimination Principle is false; but that, even if it were true, it would not help to substantiate Direct Scepticism.

2.13 A Revised Interpretation of the Sceptic's Basic Epistemic Maxim: The Contrary Prerequisite and the Defeater Prerequisite Elimination Principles

We could sum up the discussion thus far by saying that neither the Contrary Consequence nor the Defeater Consequence Elimination Principle provides a basis for Direct Scepticism. But I fear that our task is very far from complete. For there is a way of construing the claim supposedly captured by those principles in a somewhat different manner, which, if correct, would indeed supply the sceptic with the necessary grounds for Direct Scepticism. I have already alluded to this interpretation at the end of section 2.10 and I believe it may be what the sceptic really intends to be the argument for Direct Scepticism. Recall that the informal statement of the sceptic's Basic

Epistemic Maxim was: *S must be justified in rejecting H (H_c or H_d) in order to be justified in believing that p.* The two principles considered so far interpreted the "in order to be" in this basic maxim to mean that being justified in rejecting H (H_c or H_d) is a logically necessary condition of being justified in believing that p. We have seen that, interpreted in this fashion, the Basic Epistemic Maxim is not useful in providing a reason for Direct Scepticism.

Suppose, instead, that the sceptic meant to be calling attention to an *evidential prerequisite* for the justification of p, rather than merely a logically necessary condition of the justification of p. Let me clarify this way of understanding the Basic Epistemic Maxim by comparing two claims:

(1) In order to be a bachelor, a person must be a man.

(2) In order to count to ten, a person must count to five.

Claim (1) merely asserts that being a man is a logically necessary condition of being a bachelor. Whereas (2) can be understood as asserting that a person must count to five *before* counting to ten. There is no temporal priority implied by (1), whereas there is a temporal priority implied by (2).

Another example more closely related to the issues which we have been discussing may help to clarify the distinction between the Contrary *Consequence* and the Contrary *Prerequisite* Elimination Principle. Consider the following instance of the basic maxim: In order for S to be justified in believing that the pen is blue, S must be justified in believing that it is not green. Now that sentence could be taken to mean either:

(a) If S is justified in believing that the pen is blue, S is justified in believing that it is not green.

(b) S must be justified in believing that the pen is not green before S is is justified in believing that the pen is blue.

The claim embodied in (a) is merely a particular instance of the Contrary Consequence Elimination Principle. In our terminology, (a) requires that the denial of every contrary of a proposition, say x, be on the confirming chain of propositions which includes x. That requirement can be fulfilled if the denials of the contraries follow x on the chain. In fact, I argued that if x is on such a chain, then the denial of the contraries must, at least, follow x. But (b) is an instantiation of a much stronger requirement. That stronger general principle requires that S eliminate the contraries of x before S is justified in believing that x. In our terminology, that would amount to requiring that the denial of every contrary of x precede x on the chain of confirming propositions. Thus, (b) is an instance of a general epistemic principle which not only posits a logically necessary connection between x and the denial of the contraries of x, it states an evidential prerequisite.

The sceptic may have been claiming that there is an evidential priority depicted in the Basic Epistemic Maxim which is not captured by either of the interpretations considered up to this point. That is, the basic maxim might have meant that it is necessary to have evidence justifying the rejection of H (H_c or H_d) prior to being justified in accepting p. Construed in this way, Descartes' argument could be understood as involving the claim that, until he has shown that God is epistemically benevolent, he does not possess beliefs sufficiently justified to count as knowledge. And there certainly is that way of reading Lehrer's argument cited earlier (in section 2.1).

Reconsider these passages in Lehrer's account:

> Now it is not at all difficult to conceive of some hypothesis that would yield the conclusion that beliefs of the kind in question are not justified, indeed, which if true would justify us in concluding that the beliefs in question were more often false than true. The sceptical hypothesis might run as follows. There are a group of creatures in another galaxy, call them Googols, whose intellectual capacity is 10^{100} that of man, and who amuse themselves by sending out a peculiar kind of wave that affects our brain in such a way that our beliefs about the world are mostly incorrect. This form of error infects beliefs of every kind, but most of our beliefs, though erroneous, are nevertheless very nearly correct. This allows us to survive and manipulate our environment. However, whether any belief of any man is correct or even nearly correct depends entirely on the whimsy of some Googol rather than on the capacities and faculties of the man. . . . I shall refer to this hypothesis as the *sceptical hypothesis*. On such a hypothesis our beliefs about our conscious states, what we perceive by our senses, or recall from memory, are more often erroneous than correct. Such a sceptical hypothesis as this would, the sceptic argues, entail that the beliefs in question are not completely justified. . . .
>
> In philosophy a different principle of agnoiology is appropriate, to wit, that no hypothesis should be rejected as unjustified without argument against it. Consequently, if the sceptic puts forth a hypothesis inconsistent

with the hypothesis of common sense, then there is no burden of proof on either side, but neither may one side to the dispute be judged unjustified in believing his hypothesis unless an argument is produced to show that this is so. If contradictory hypotheses are put forth without reason being given to show that one side is correct and the other in error, then neither party may be fairly stigmatized as unjustified. However, if a belief is completely justified, then those with which it conflicts are unjustified. Therefore, *if neither of the conflicting hypothesis is shown to be unjustified, then we must refrain from concluding that belief in one of the hypotheses is completely justified.* [Emphasis added.] . . . Thus, *before* scepticism may be rejected as unjustified, some argument must be given to show that the infamous hypotheses employed by sceptics are incorrect and the beliefs of common sense have the truth on their side. [Emphasis added.] If this is not done, then the beliefs of common sense are not completely justified, *because conflicting sceptical hypotheses have not been shown to be unjustified.* [Emphasis added.] From this premiss it follows in a single step that we do not know those beliefs to be true because they are not completely justified. And then the sceptic wins the day.

In those passages Lehrer appears to be claiming that the evidence, e, which justifies S in believing that p must contain the denial of H, since "conflicting sceptical hypotheses" must be shown to be unjustified "before" p is acceptable. If we use the notion of "grounded belief" introduced during the defense of the Contrary Consequence Elimination Principle, what the sceptic is now claiming is that ∼ H must precede p in the chain of grounded beliefs. That is, to say that the evidence for p "contains" the proposition ∼ H is to say that ∼ H precedes p in the chain of grounded beliefs. As Lehrer puts it in the passage I quoted, "If this [i.e., giving an argument which shows that H is incorrect] is not done, then the beliefs of common sense are not completely justified, because conflicting sceptical hypotheses have not been shown to be unjustified." Hence, the epistemic principles employed by the sceptic in this interpretation of the Basic Epistemic Maxim are as follows:

Contrary Prerequisite Elimination Principle	For any propositions, x and y, (necessarily) if y is a contrary of x, then if e is adequate confirming evidence for x, then e contains ∼ y.
Defeater Prerequisite Elimination Principle	For any propositions, x and y, (necessarily) if y is a defeater of the justification for x, then, if e is adequate confirming evidence for x, then e contains ∼ y.

To make these interpretations of the Basic Epistemic Maxim more intuitively clear, imagine a variety of sets of evidence, e_1 . . . e_n, each *apparently* providing adequate confirming evidence for p. (I say "apparently" in order not to beg the issue against the sceptic.) For example, let p be 'The table in my living room is brown', and let the following be sets of apparently adequate confirming evidence:

e_1 The table looks brown to me, and nothing seems odd about the perceptual situation. Other people agree that it is brown.

e_2 Although I cannot now see the table, I bought a brown table this morning and the furniture store where I purchased it confirms the delivery of the table. The person with whom I spoke said it was placed next to a green chair in my living room. I do have such a chair in my living room.

e_3 If the table is brown, it will affect the spectroscope in a particular way, and it does.

One could imagine additional sets of seemingly adequate confirming evidence, or could require that e_1 through e_3 be strengthened, but the point of the sceptic's Contrary Prerequisite Elimination Principle would be to require that every evidence set, e, sufficient to confirm p must include the denial of every contrary of p. Specifically, e must contain 'it is not the case that the table is green' and 'it is not the case that (the table is not brown and everyone has suddenly acquired a strange form of color blindness)' and 'it is not the case that (the table is not brown but looks brown because of some peculiar lighting in my room)', etc.

To return to the sceptical hypothesis H_c, the Contrary Prerequisite Elimination Principle would require that *before* S can be justified in believing that p, S must have adequate confirming evidence that $\sim H_c$. In our terminology, this principle requires that $\sim H_c$ precede p in the chain of grounded beliefs for S. The argument for Direct Scepticism would continue by asserting that S has no such adequate confirming evidence against H_c and, therefore, S is not justified in believing that p.

2.14 The Evaluation of the Contrary Prerequisite and the Defeater Prerequisite Elimination Principles: Unacceptable

Now what are we to make of these interpretations of the sceptic's Basic Epistemic Maxim? In order to answer that question it is important to keep in mind the fundamental difference between the consequence and prerequisite principles. Although the Contrary Consequence Elimination Principle asserts that if Jsp, then S is justified in denying every contrary of p, it did not require that the evidence for p contain the denials of all the contraries of p. The contrary Prerequisite Elimination Principle is considerably stronger—for, according to it, S must *first* eliminate all the contraries of p in order to be justified in believing that p.

There is another important difference between the consequence and prerequisite formulations of the sceptical epistemic maxim. For our task here is somewhat easier than it was in considering the two Consequence Principles. Since the Defeater Prerequisite Elimination Principle entails the Contrary Prerequisite Elimination Principle, if it can be shown that the Contrary Prerequisite Elimination Principle is false, we will have disposed of both. The entailment between the two Consequence Principles obtains as well, but I argued that the Contrary Consequence Elimination Principle was true. So that strategy was not available to us before. The entailment between both sets of principles holds because every logical contrary of p is a defeater of the justification of p.

The strategy here will be to show that the Contrary *Prerequisite* Elimination Principle is false in order to show that both Prerequisite principles are false. Note first that the considerations which led us to accept the Contrary *Consequence* Elimination Principle are not transferable to the prerequisite principle. Roughly, the Contrary Consequence Elimination Principle was correct because whenever S had adequate confirming evidence for p, S had adequate confirming evidence, namely p itself, for the denial of every contrary of p. I argued that p preceded the denial of the contraries, for example $\sim H_c$, in the chain of S's grounded beliefs. That is, p provided the confirmation of $\sim H_c$; or, put another way, the evidence for $\sim H_c$ contained p. On the other hand, the Contrary Prerequisite Elimination Principle requires that the denial of every contrary of p, e.g., $\sim H_c$, be part of the evidence for p. Thus, it requires that $\sim H_c$ precede p in the chain of grounded beliefs. Hence, the argument for the Contrary Consequence Elimination Principle cannot be transferred here; for that argument depended upon assuming that p was already part of the grounded evidential chain and concluded by showing that $\sim H_c$ followed it in the chain. An argument supporting the prerequisite principle could not begin with the assumption that p is part of the chain of grounded beliefs unless the sceptic is willing to grant that $\sim H_c$ was already present in the chain of grounded beliefs. And, of course, the sceptic is not prepared to grant that, since that would be to grant that S has adequate evidence to reject the sceptical hypotheses.

So are there other considerations for believing that in order to be justified in believing that p, S must first eliminate all the contraries of p? Well, there are occasions when we arrive at a belief that p by, among other things, eliminating some of the contraries of p. However, the prerequisite principle requires that we eliminate *all* the contraries of p. And it is that requirement which must be examined in order to determine whether it accurately reflects the requirements of justification.

In the passage quoted earlier, Lehrer may have already indicated that the Contrary Prerequisite Elimination Principle requires more of S than what is ordinarily required. For in those paragraphs he contrasts the principles of justification used in philosophy with those ordinarily used. And, at least with regard to our ordinary practices, he is correct! Of course, we do not ordinarily insist that every contrary of p be eliminated before we are willing to grant that p is justified.

Consider any of the stereotypic detective stories. At the end, when the clever detective summarizes the evidence leading to the identification and arrest of the culprit, he/she does not include in the summary of the case the denial of all the contraries to the claim that the culprit did it. Imagine how many that would be! For example, must the detective in every story first eliminate the possibility that Queen Elizabeth in a clever disguise did it? Does he/she first have to eliminate the possibility that the Ghost of Christmas Past is real and that the Ghost did it just to show that all is not well with the world? Of course not. When twenty-five contraries were eliminated, would that be enough? How about fifty? Would the detective story ever end if this were a requirement? The point is that the story does end, and we find that plausible.

The principle does not accurately characterize our ordinary practices, but, as mentioned above, I suspect that the sceptic may readily grant that and argue that our ordinary practices are inadequate to bring about knowledge. The sceptic could suggest, and indeed it appears that Lehrer was suggesting, that although the Contrary

Prerequisite Elimination Principle does not accurately characterize our ordinary practices, for some purposes a more stringent set of requirements is necessary. After all, it might be claimed, "in philosophy" or in some other context where certainty is the goal, we ought to treat all hypotheses equally and be willing to investigate *all* the alternatives to those normally accepted.

There are two responses to this suggestion. First, it would seem that if the sceptic recognizes that this suggested principle of justification requires more than what is ordinarily required, he/she has tacitly agreed that this cannot provide a reason for accepting Direct Scepticism. For since the requirements of "justification" and, hence, "knowledge" as well, have been changed drastically by the sceptic, the sceptic is not denying that we have what we formerly believed was knowledge.

Put another way, if the sceptic asserts that we do not have knowledge, and the nonsceptic asserts that we do have knowledge, the disagreement can be of two sorts. The sceptic and the nonsceptic can agree about the nature of the necessary conditions of knowledge and differ over whether those conditions can be fulfilled; or the sceptic and the nonsceptic can be disagreeing over the nature of the necessary conditions themselves. I said in section 1.1 that the Moorean response to the sceptical challenge (roughly, "We have knowledge, so scepticism is false") was philosophically unacceptable because scepticism as usually formulated seemed to emerge from within the concepts of knowledge and justification shared by both the sceptic and the nonsceptic. The sceptic appeared to be appealing to universally accepted necessary conditions of epistemic justification. On the other hand, if the sceptic were to insist that in order for S to be justified in believing that p, S must first eliminate all the contraries of p, the obvious result would be Direct Scepticism. But the Moorean response gains credibility as the requirements of justification and knowledge become increasingly more stringent than those universally accepted. For whatever the alleged value gained by adopting the more rigorous requirements, the cost (i.e., the loss of knowledge) is too high. And, as I have said, the dispute with scepticism appeared to be over the scope of knowledge, not this newly created, so-called "knowledge."

Be that as it may, however, there is a second objection to this suggestion which I believe shows that the scpetic has violated one of the initial ground rules of the dispute. I believe it can be shown that the Contrary Prerequisite Elimination Principle actually requires that in order for e to be adequate evidence for p, e must entail p.

Consider the set of *all* the contraries of p: C_1, C_2, C_3 ... C_n. The Contrary Prerequisite Elimination Principle requires that $\sim C_1$ be included in e, and that $\sim C_2$ be included in e, etc. In fact the principle requires that e contain ($\sim C_1$ & $\sim C_2$ & C_3 ... $\sim C_n$), since e must contain the denial of all the contraries of p. Now e entails p, if (e&$\sim$p) is false in every possible world. But (e&$\sim$p) is false in every possible world since $[(\sim C_1 \& \sim C_2 \& \ldots \sim C_n)\&\sim p]$ is false in every possible world.

To show that, consider any world, w_i, in which $\sim$p and some other proposition, call it 'r', is true. In that world there is a true contrary of p, C_i, namely ($\sim$p&r). Hence, in that world $\sim C_i$ is false. Since the conjunction of the denial of *all* contraries of p includes $\sim C_i$ (and $\sim C_i$ is false), that conjunction is false. Thus it is not possible for $\sim$p to be true and the conjunction of the denial of all of the contraries of p to be true. Thus ($\sim C_1 \& \sim C_2 \& \sim C_3$ ... $\sim C_n$) entails p. Since e includes that conjunction (by the Contrary Prerequisite Elimination Principle), e entails p.[35]

There is a more simple argument with the same consequence which does not make use of possible worlds. The propositions '(q&$\sim$p)' and '($\sim$q&$\sim$p)' are both contraries of p. Hence, they would both be included in the set of all the contraries of p. But since the proposition '$\sim$(q&$\sim$p) & $\sim$($\sim$q&$\sim$p)' entails p, any conjunction containing that conjunction will entail p. Thus any conjunction containing the denials of all the contraries of p will entail p.

Hence, I conclude that whatever reasons prompt the sceptic to adopt the Contrary Prerequisite Elimination Principle also force the sceptic to require that the evidence, e, for a proposition p, entail p. But that consequence is sufficient to reject that principle as providing a basis for Direct Scepticism, since as we have seen, the sceptic cannot require that e entail p (see section 1.4).

2.15 Other Unsuccessful Versions of the Contrary Prerequisite Elimination Principle

In the previous section, we saw that the Contrary Prerequisite Elimination Principle has the unacceptable consequence that if e is adequate confirming evidence for p, then e entails p. The sceptic, however, may reply that the Contrary Prerequisite Elimination Principle is stronger than he/she requires in order to capture the informally stated Basic Epistemic Maxim that S must be justified in rejecting H_c prior to being justified in believing p. The fact that the Contrary Prerequisite Elimination Principle results in e entailing p appears to obtain because the principle requires that e contain the denials of *all* the contraries of p. The sceptic may suggest that S does not need to reject every contrary of p. For example, S need not reject $(\sim e \& \sim p)$ since that contrary is not compatible with the original evidence, e. Thus the sceptic may, quite reasonably, reduce what is required by the Contrary Prerequisite Elimination Principle to the weaker claim that only those contraries compatible with the original evidence need to be eliminated. Let us call that original evidence, e^1, and the compatible contraries, C_1^1, C_2^1, . . . C_n^1. It may be hoped that such a restricted version of the principle does not lead to the unwarranted entailment.

But, unfortunately for the sceptic, this restricted version of the principle leads to the same result. For presumably we could find a contrary of p, say $(q \& \sim p)$ which is such that it and the contrary $(\sim q \& \sim p)$ are both compatible with e^1. Hence, the set of compatible contraries would include both of those contraries, and the conjunction of the denials of all contraries compatible with e_1 will, once again, entail p. Thus, this revised version of the Contrary Prerequisite Elimination Principle does not avoid the violation of the ground rules of the dispute.

There is, however, one further variation of this principle which the sceptic may wish to explore. The sceptic may wish to require, not that every contrary of p compatible with e need be eliminated, but merely that H_c be included in the set of those which must be eliminated before S is justified in believing that p. That is, it may be claimed that there is something unique about the sceptical hypothesis, H_c. It could be argued that H_c is one member of a set of contraries of p each of which would call into question not only the justification of p, but the justification of all empirical contingent propositions arrived at by nondeductive inference. For H_c challenges the reliability of any such principle of confirmation. Hence, the sceptic could assert that before we can be justified in believing that p (or any other proposition based upon nondeductive inference), we must eliminate the possibility that there are no valid nondeductive principles of confirmation whatsoever.

The fact that we ordinarily believe that some empirical propositions are confirmed could be acknowledged by the sceptic. But it could be claimed that this is irrelevant since the argument here is that, appearances to the contrary, no empirical proposition that p is justified until the principles of confirmation used to arrive at p have been "vindicated." We may have the conviction that we are justified in believing that p, but we are not so justified until H_c is shown to be false.

Once again the sceptic would claim that we are not justified in rejecting H_c; and hence we are not justified in believing that p. But this argument, which requires that e contain $\sim H_c$, is one which, as I indicated in section 1.2, confuses those considerations which would be appropriate for a defense of Iterative Scepticism with those which would lead to Direct Scepticism. Direct Scepticism asserts that we cannot know that p; whereas, Iterative Scepticism asserts that, although we may know that p, we cannot know that we know that p. The argument just considered which claims a unique status for the Contrary Prerequisite Elimination Principle amounts to asserting that before we can be justified in believing that p we must first show that the principles of confirmation which lead us to p are "valid." But, as I argued, in order to be justified in believing that p, all that is required is that the appropriate epistemic principle be valid; it is not further required that we be able to show that it is valid. That is, knowledge that p on the basis of e requires that eCp, but it does not further require that we have adequate confirming evidence for the claim that eCp. Thus, if this is an argument for anything at all, it is that in order to know that we can know that p, it is required that we show that H_c is false. Now whether this argument is sound is beside the point; for it is not an argument for Direct Scepticism, but rather one for Iterative Scepticism.

2.16 A Seemingly More Moderate Sceptical Epistemic Principle

There is one remaining point to be considered before concluding my argument to show that the sceptic has failed to present a good reason for accepting Direct Scepticism. At the beginning of the discussion of the Evil Genius Argument, I pointed out that although my strategy employed a version of the paradigm case argument and such an argument is not conclusive against Direct Scepticism, the strategy employed here may nevertheless prove effective. For other arguments which are not apparently dependent upon the Basic Epistemic Maxim underlying the Evil Genius Argument may share the defects of the Contrary Consequence Elimination Principle and the Contrary Prerequisite Elimination Principle. The following example will illustrate what I meant.

Suppose that a defender of Direct Scepticism were to propose a principle apparently more moderate than either the Contrary Consequence or Contrary Prerequisite Elimination Principle. Call it the *Moderate Sceptical Epistemic Principle.*

Moderate Sceptical Epistemic Principle	If S is justified in believing that p on the basis of e (eJsp), then e must be adequate confirming evidence for p; and the latter requires that e make p epistemically preferable to any contrary of p, say q.

I say *epistemically preferable* since I assume that there may be good reasons other than epistemic ones for accepting p rather than q; and these reasons are not relevant to the discussion here. (Pascal's wager might be considered such a case.) The sceptic would then claim that the evidence for p does not make it epistemically preferable to H_c.

As mentioned above, the Moderate Sceptical Epistemic Principle appears to be a less stringent proposal than any of the principles which motivate the Evil Genius Argument. After all, it does not appear that the sceptic is requiring that e be sufficient by itself to justify ~q; it must merely make p epistemically preferable to q. Also, the sceptic does not appear to be requiring that e contain ~q. Consequently, it appears that the Moderate Sceptical Epistemic Principle may provide the reason for Direct Scepticism sought by the sceptic. But on closer inspection we will see that this principle is vulnerable to the same criticisms we brought against the other versions of the basic maxim. All the considerations which led to the conclusion that the Contrary Consequence and Contrary Prerequisite Elimination Principles do not provide a reason for accepting Direct Scepticism apply equally well to the Moderate Sceptical Epistemic Principle.

To see that, notice that, like the Contrary Consequence Elimination

Principle, this principle follows from the characterization of justification developed in section 2.7, namely, if S is justified in believing that p, then there is a grounded proposition, w, for S such that w provides adequate confirming evidence for p, *and* there is no other grounded or pseudogrounded proposition, u, for S which overrides w for p. The Moderate Sceptical Epistemic Principle specifies that the grounded adequate confirming evidence is e. Thus, if eJsp, then eCp (because e is the instantiation of w); but also notice that if e were to provide adequate confirming evidence for some contrary of p, say q, then there would be a grounded overriding proposition for S, namely, q— since (e&q)Cp. So e cannot be adequate evidence for q. Hence, if eJsp, then eCp and eCq. And, finally, if eCp and eCq, then e makes p epistemically preferable to q. Thus, the same considerations which provide support for the Contrary Consequence Elimination Principle can be transferred to support the Moderate Sceptical Epistemic Principle.

But once this analogy is noticed, another should become obvious—namely, the Moderate Sceptical Epistemic Principle (thus interpreted) provides no warrant for Direct Scepticism. For, presumably, the nonsceptic would believe that e was adequate confirming evidence for p and not adequate confirming evidence for H_c and would ask the sceptic to provide a *reason* for thinking otherwise. That is, the nonsceptic need not (at this stage of the argument—the stage contained in Chapter Two) provide a reason for thinking that p is epistemically preferable to H_c; it is the sceptic who must provide a reason for thinking that either e is not adequate confirming evidence for p or that e is adequate confirming evidence for H_c, if the Moderate Sceptical Epistemic Principle is to provide the sceptic with a reason for asserting Direct Scepticism. I say that the burden of proof is on the sceptic because this chapter is devoted to an examination of the reasons the sceptic might have *for* Direct Scepticism. The burden shifts to my shoulders in Chapter Three.

I doubt that the sceptic will choose to argue that e is adequate evidence for H_c, since he/she would then be put in the awkward position of claiming that S can know that H_c. In other words, the sceptic would be arguing that although we cannot know that Jones owns a Ford, we can know that there is an evil genius making us falsely believe that Jones owns a Ford!

So the sceptic must choose the other path; namely, the path which shows that e is not adequate confirming evidence for p. Now, what reason could the sceptic advance for believing that e is not adequate confirming evidence for p? Well, the sceptic could say: e is not adequate evidence for p because it is equally good evidence for p and for H_c and, in order to be adequate confirming evidence for p, it must make p epistemically preferable to H_c. But that "argument" is too tight a circle to be useful, since if the sceptic is to show that e does not make p epistemically preferable to H_c, he/she would have to show that it is not the case that eCp and $e \not C \sim H_c$.

However, I suspect that the sceptic would take one of the following four alternatives in order to show that e is not adequate confirming evidence for p:

a_1 e is not adequate evidence for p because, if it were, S would be justified in denying H_c, and S is not justified in denying H_c.

a_2 e is not adequate evidence for p because, if it were, S would be justified in denying all the defeaters of the justification of p, and S is not justified in denying H_d.

a_3 e is not adequate evidence for p because, if it were, e would contain the denial of H_c, and e does not contain $\sim H_c$.

a_4 e is not adequate evidence for p because, if it were, e would contain the denial of H_d, and e does not contain $\sim H_d$.

But these reasons are nothing more (nor less) than appeals to the Contrary Consequence Elimination Principle, the Defeater Consequence Elimination Principle, the Contrary Prerequisite Elimination Principle, and the Defeater Prerequisite Elimination Principle, respectively. And thus, all the considerations relevant to those principles are applicable here. Far from the "Moderate" Sceptical Epistemic Principle being more moderate than the other epistemic principles which we have already considered, it appears to depend upon them. The seemingly more moderate principle is only a disguised appeal to the very principles which we found were unable to provide the sceptic with a reason for accepting Direct Scepticism.

2.17 Summary of Chapter Two

Let me sum up the argument in Chapter Two. The sceptic advanced several interpretations of his/her Basic Epistemic Maxim — the maxim motivating the Evil Genius Argument. That basic maxim was: In order for S to be justified in believing that p, S must be justified in denying the Evil Genius Hypothesis. We saw that there were two distinct ways of interpreting the Evil Genius Hypothesis. One interpretation represented the hypothesis as a logical contrary of p; whereas the other made it a defeater of the justification for p. We also saw that there were two ways of understanding the expression 'in order to be' in the basic maxim. In one sense, the grammatical consequent in the statement of the basic maxim was understood to be a mere logical consequent of the grammatical antecedent; in the other sense, the grammatical consequent was taken to be stating an evidential prerequisite of the grammatical antecedent. Thus, we arrived at four interpretations of the sceptic's Basic Epistemic Maxim. No interpretation of the maxim was found both to state a valid epistemic principle and to motivate Direct Scepticism.

The Contrary Consequence Elimination Principle

This principle asserted that if S is justified in believing that p (where 'p' represents a proposition believed to be knowable), then S is justified in denying H_c, where 'H_c' stands for a contrary of p such as: e (the evidence for p) &~p & the evil genius causes the false belief that p. I argued that this principle was correct by showing that (1) the objections to it were inconclusive and that (2) it could be derived from an acceptable general model of justification. But even though the principle is correct, I argued, it did not lead to scepticism because, in order for the sceptic to show that S failed to possess a justification for $\sim H_c$, the sceptic would have to provide a good reason for believing that no sufficient condition of that justification obtained. Since being justified in believing that p is *the* sufficient condition of being justified in denying that H_c, which the nonsceptic believes does obtain, the sceptic would have to provide a good reason for believing that S was not justified in believing that p. That, however, was the purported conclusion of the sceptical argument. Thus, the nonsceptic would find the Contrary Consequence Elimination Principle equally useful for demonstrating that S was justified in denying that $\sim H_c$. I argued that it was neither irrational, epistemically foolhardy, nor dogmatic for the nonsceptic to claim that S was justified in denying that H_c.

The Defeater Consequence Elimination Principle

This principle asserts that if S is justified in believing that p, then S is justified in denying that H_d (where 'H_d' stands for the conjunction: e & there is an evil genius which could bring it about that p is false and S believes that p). I argued that this principle was invalid by showing that (1) there were counterexamples to it and (2) it was not derivable from the general model of justification used to support the Contrary Consequence Elimination Principle. Further, even if the principle were valid, it could not provide a basis for Direct Scepticism for the same reasons which showed that the Contrary Consequence Elimination Principle failed to motivate Direct Scepticism.

The Contrary Prerequisite and the Defeater Prerequisite Elimination Principles

The Contrary Prerequisite Elimination Principle took the sceptic's Basic Epistemic Maxim to be stating an evidential prerequisite for justification rather than merely a logical consequence of justification. It requires that in order for S's evidence for p to be adequate confirming evidence, it must include the denials of all the logical contraries of p. According to it, S must first eliminate the contraries of p before being justified in believing that p. This principle follows from (i.e., is a corollary of) the stronger principle—the Defeater Prerequisite Elimination Principle—which requires that the adequate confirming evidence for p include the denials of all the defeaters of the justification for p. I argued that both principles were invalid. As they stood, both failed to portray accurately the necessary conditions of justification. In addition, they had the consequence that the sceptic was tacitly requiring that if eCp, then e entails p. In section 1.4, I argued that such a requirement robbed scepticism of its initial interest.

We examined various modifications of these four principles in order to determine whether any of them could be made to be acceptable while still motivating Direct Scepticism. I argued that no revisions of the principles could satisfy both requirements.

Thus, the argument in the second chapter could be stated briefly as follows: The sceptic's Basic Epistemic Maxim can be interpreted in a variety of ways. Each interpretation of the maxim supposedly identifies an unfulfillable necessary condition of justification, which is itself a necessary condition of knowledge. My arguments were designed to show that each principle was either:

(1)   such that it provides no warrant for Direct Scepticism, or

(2)   such that it requires that e entail p, or

(3)   such that it conflates the distinction between Direct Scepticism
       and Iterative Scepticism.

Thus, I conclude that the Evil Genius Argument does not provide us with a reason for believing that S never knows that p.

2.18 A Sceptical Rejoinder and the Task Remaining for Chapter Three

Thus Far the argument has been that the only interpretation of the Evil Genius Argument which could provide a warrant for Direct Scepticism depends upon an epistemic principle of justification so strong that it requires that the evidence for a proposition entail that proposition.

But suppose that the sceptic accepts the conclusion of the arguments in Chapter Two and replies:

> If Direct Scepticism is committed to the claim that S knows that p only if S's evidence for p entails p, so be it. Perhaps that had not been recognized before, but far from that constituting an argument *against* scepticism, what has been revealed is one of the essential features *of* scepticism.

> The argument throughout this chapter has presupposed that justifications can be defeated by further evidence. That is, it has been presumed that justifications are, *in principle*, defeasible. If that were the case, then no belief could ever be certain. But knowledge entails certainty. Thus, the stringent condition that e must entail p is not only essential to scepticism, but is, in addition, required by a proper analysis of knowledge.

The sceptic's rejoinder has three parts. First, he/she is now forsaking what I called, in Chapter One, the historical context of the argument, one ingredient of which was that the evidence, e, for an empirical, contingent proposition, p, need not entail p in order to be adequate to justify it. Second, he/she is asserting that knowledge entails certainty. And finally, he/she is supposing that only the stringent justification condition that e entail p can plausibly account for the claim that knowledge entails certainty.

What can be said in response to these three claims? First, as I argued in section 1.4, if the sceptic endorses the first point, the position becomes relatively uninteresting, since that interest depends upon the apparent emergence of scepticism within the context of the justification of empirical, contingent propositions. That type of justification is such that having adequate confirming evidence for some proposition does not require that there be no additional, defeating evidence against it. Thus, if the sceptic endorses either the Contrary Prerequisite or the Defeater Prerequisite Elimination Principle, he/she is not denying what is ordinarily meant when we assert that S knows that p.

Further, if it is required that e entail p, the "arguments" presented for scepticism based upon the sceptical hypothesis become seriously misleading. For there would then be an extremely short (and conclusive) argument for scepticism:

(1) If Ksp, then every link in S's evidential chain terminating in p is an entailment.

(2) There are no such evidential chains.

∴ ~ Ksp.

The second step, (2), can be granted (at least with respect to empirical knowledge); but if (1) accurately portrayed justification, the sceptical hypothesis would be beside the point. The mere fact that (e&~p) is not a self-contradiction (a fact readily granted by the defender of empirical knowledge) would be, along with (1), sufficient to establish Direct Scepticism.

The Evil Genius Argument becomes seriously misleading if it is required that e entail p. For implicit in that argument is the assumption that the fact that e failed to entail p is not enough, by itself, to warrant Direct Scepticism. If it were, there would be no point at all in developing the Evil Genius Argument! The argument proceeded, however, by attempting to show that, for one reason or another, e was not adequate confirming evidence for p. But if (1) is what the sceptic really means to be asserting, the introduction of the various epistemic principles like the Contrary Consequence or Contrary Prerequisite Elimination Principles is misleading. For Direct Scepticism would follow immediately from the proposition that there are no evidential chains in which every link is an entailment.

The second and third parts of the sceptic's reply, however, are worth considering. It is alleged that knowledge entails certainty and that it is not possible for a proposition to be certain if the evidence for it is, *in principle*, defeasible and falls short of entailing it. As I stated at the beginning of the discussion in Chapter One, I am willing to grant that knowledge does entail absolute certainty. Thus, Chapter Three is designed to sketch an adequate account of certainty which is both acceptable to the sceptic and within the scope of a defeasibility analysis of knowledge.

CHAPTER 3

ABSOLUTE CERTAINTY
IN THIS WORLD

3.1 Review of the General Argument

It seems best to begin this chapter by reviewing the role it is to play in the proposed refutation of scepticism. Briefly, the aim here is to delineate an account of absolute certainty which both captures all the intuitively plausible features of the sceptic's insistent claim that knowledge entails absolute certainty and is such that many "ordinary," empirical contingent propositions are, in fact, absolutely certain.

In order to more fully understand the importance of the role of this chapter, let me restate the overall argument of this book, indicating what I believe has already been accomplished and what remains to be done.

The three forms of scepticism which this book is designed to refute can be identified by indicating their varied responses to the question: Does S know that p (where 'p' is taken to represent a proposition normally believed to be knowable)? Direct Scepticism asserts that Ksp cannot be true, since one of the necessary conditions of knowledge cannot be fulfilled. Iterative Scepticism asserts that Ksp may be true but that it cannot be known. Finally, Pyrrhonian Direct Scepticism asserts that there are equally good (or bad) reasons both for Ksp and against Ksp.

These forms of scepticism can be refuted if it can be shown that:

(1) There is no good reason for believing that Ksp is always false.

(2) There is no good reason for believing that Ksp is sometimes true.

In Chapter Two I argued for (1) by examining what I took to be the epistemic principles used by the sceptic in his/her argument for Direct Scepticism. We considered four such principles and many variations in order to show that within the ground rules of the debate discussed in Chapter One, no principle was both acceptable and capable of providing evidence for the claim that S cannot know p. The most important of the ground rules for that argument was that the sceptic cannot insist that in order for evidence, e, to be adequate to confirm an empirical, contingent proposition, p, it must entail p. That is, the sceptic cannot insist that if eCp, then e → p. I argued that the only epistemic principles capable of motivating scepticism forced the sceptic to claim that if eCp, then e → p.

At this point, the sceptic will, no doubt, remind me that there were two components of the ground rules delineated in the first chapter. In addition to the one mentioned above, which played a crucial role in the argument of the second chapter, I had granted to the sceptic the right to insist that knowledge entails absolute certainty. The sceptic will repeat the claim made in his/her rejoinder at the end of Chapter Two—namely, if a proposition, p, is rendered absolutely certain for S by some evidence, e, then e must entail p. Other conditions may be necessary as well; for example, e must be known by S, and the entailment between e and p must also be known by S. But what is crucial to note, according to the sceptic, is that it is possible to satisfy the condition that knowledge that p entails absolute certainty that p only if our evidence, e, for p guarantees the truth of p, and *that* is possible only if e entails p.

The task of this chapter is to show that e can render p absolutely certain without entailing p. Specifically, I will develop and defend an account of knowledge which has the consequence that if S knows that p on the basis of e, then e renders p absolutely certain for S, but which is such that many contingent, empirical propositions normally believed to be knowable are in fact known. If this account of knowledge, or one sufficiently similar to it, is correct and if we have good reasons for believing that *all* the necessary and sufficient conditions of knowledge are jointly fulfilled, then we will have located the good reasons for believing that S sometimes knows that p, mentioned above in (2). In other words, the function of this chapter is to defend (2).

3.2 Some Preliminary Remarks About Certainty

As mentioned in the previous section, the task of the third chapter is to provide an account of knowledge which has as one consequence that S knows that p only if p is absolutely certain for S and, as a second consequence, that many empirical, contingent propositions are in fact known. In this section, I will briefly consider some alternative approaches which philosophers have taken to the analysis of the concept of certainty in order to clarify my own approach by contrasting it with those alternatives.

Let me begin by saying what it is that I will *not* be claiming. I will not try to show that we would *ordinarily attribute* knowledge that p to S only when p is absolutely certain for S. Some philosophers believe that we do not customarily reserve the attribution of 'knowledge' to those propositions for which 'certainty' would also be appropriate. They would hold that many propositions are believed to be known which are not believed to be absolutely certain. I will discuss that briefly at the end of this section and consider it in more detail in sections 3.5 and 3.6.

However, there is another view, attributable to Wittgenstein in his *On Certainty*,[1] which I would like to consider here. If I understand him correctly, he claims that we customarily attribute knowledge that p to a person only when p is *not* certain for that person. Consider the following excerpts:

308. 'Knowledge' and 'certainty' belong to different *categories*. They are not two 'mental states' like, say 'surmising' and 'being sure'. ...

504. Whether I *know* something depends on whether the evidence backs me up or contradicts me.

445. But if I say "I have two hands", what can I add to indicate reliability? At the most that the circumstances are the ordinary ones.

446. But why *am* I so certain that this is my hand? Doesn't the whole language-game rest on this kind of certainty?

Or: isn't this 'certainty' already presupposed in the language-game? Namely by virtue of the fact that one is not playing the game, or is

playing it wrong, if one does not recognize objects with certainty.
282. I cannot say that I have good ground for the opinion that cats
do not grow on trees or that I had a father and a mother.

Wittgenstein's point seems to be that to claim ⌐p is certain⌐ is to
indicate that no grounds for p are available.[2] For it is to assert
that p has the status of a proposition in the linguistic practices of
the speaker's community such that its acceptability does not depend upon *evidence* for its truth but rather that its acceptability
is inherent in the linguistic practices themselves. For example,
there are *no* circumstances in which evidence can be presented for
the proposition 'there are physical objects'—at least within *our*
"language game." Moore's famous claim—'there is one hand'—is a
proposition which is certain in the *ordinary* circumstances (i.e.,
holding the hand right before your eyes in broad daylight) just because we would not know what could be grounds for its truth.

Morawetz, commenting on *On Certainty*, points to the two types
of certain propositions mentioned in the preceding paragraph in
the following:

First, there are cases in which *p* is neither a matter that can be justified
and doubted nor a matter for which corroborating resources can be
sought. For example, "There are physical objects" and "Objects continue
in existence when not being perceived" express matters that are never
matters for testing.

Second, there are cases in which *p* is not a matter that can be justified
and doubted in ordinary contexts but is a matter for which corroborating
resources can be found. I cannot doubt or seek to justify that I have two
hands or that my words mean what they do or that my name is T.M. without jeopardizing my confidence in my ability to judge at all, except in
specific, extraordinary cases. But I *can* nonetheless find my name in my
wallet, count my limbs, and check the dictionary.[3]

This passage from Morawetz underscores the important point for
our discussion here. For if we combine the view that the proposition
'here is one hand' cannot be justified (at least, in the ordinary
circumstances) with the claim in Wittgenstein's par. 243 that "One
says 'I know' when one is ready to give compelling grounds," then
Wittgenstein's reasons for believing that 'knowledge' and 'certainty'
belong to different categories become clear. To say that p is known
is to invite an inquiry concerning the evidential grounds for p, because

one is prepared to present compelling grounds. Whereas to say that p is certain is to attempt to foreclose the request for grounds. Thus, Wittgenstein thinks that Moore was correct in claiming that 'here is one hand' is certain but that he was incorrect in claiming that 'here is one hand' is known — at least, in the ordinary circumstances.[4]

Wittgenstein's view may be correct about the ordinary or customary conditions of the attribution of 'knowledge' and 'certainty'. That is, perhaps we usually do attribute knowledge that p only when we are prepared to present what we take to be compelling evidence for p and we attribute certainty to p in order to block an inquiry concerning the evidence for p. But, in spite of that, two points need to be noted.

First, from the fact that, customarily, when we attribute knowledge that p to S, we do not also attribute certainty that p to S, it does not follow that such attributions are false. As Ayer has said:

> In the ordinary way, the point of saying 'I know that so and so' is either to show that one has acquired some information which one may have been thought not to possess, or else to insist on the truth of some proposition, which has been or might be seriously challenged. In either case saying 'I know' gives an assurance which is assumed to be needed. But when the proposition in question is that I am seeing some familiar colour, or that these are my hands, there is normally no need for any such assurance. It is taken for granted that I can recognize ordinary colours: the proposition that these are my hands is not, in this context, one that anybody is seriously inclined to doubt. ...
>
> From this we can conclude not only that the use of the expression 'I know that' may serve no purpose, but that it may be actually misleading. It does not, however, follow that its use in such cases is senseless, or even that it is not used to state what is true. This may be illustrated by another example. Although people most commonly believe what they assert, the conventional effect of prefacing an assertion with the words 'I believe that' is to weaken its force. It suggests that one is not entirely sure of what follows. If I say, for instance, 'I believe that Smith has been elected' I commit myself less than if I had said outright: 'Smith has been elected'. Nevertheless, it certainly does not follow that when I do assert something outright, I do not believe it. In exactly the same way, the fact that it may be pointless or even misleading for me to say such things as 'I know that this looks red to me' or 'I know that these are my hands' in no way entails that what I am saying is not true.[5]

Wittgenstein's remarks do not constitute conclusive grounds for rejecting the entailment between knowledge and certainty as those concepts are customarily employed. For, as others have pointed out, the fact that the utterance ⌐that's not a house, it's a mansion⌐ is immediately understood and often acknowledged as correct does indicate that 'house' and 'mansion' are not normally attributed to the same set of buildings and also that, usually, it would be misleading to call a mansion an (ordinary) house.[6] But it is nevertheless not false that mansions are houses. Similarly, it may be *misleading* to call a certain proposition (i.e., a proposition which is certain) a known one, but it is nevertheless on that account not false that certain propositions are known propositions.

But there is a second, more important point to be noted. Epistemologists, especially Cartesian ones with or without sceptical inclinations, are often interested in two projects: cataloguing our knowledge and ordering the types of propositions which can be known by employing specifiable evidential relationships. Thus, these epistemologists have participated in helping to create a nonordinary context in which it is not only possible but customary to include propositions which are certain among those which are known. C. D. Rollins summarizes standard philosophic practice when, writing in the *Encyclopedia of Philosophy,* he comments that "it has become apparent that there is a connection between certainty and knowledge in that whatever is certain is known."[7] The following passages from Russell's *Inquiry Into Meaning and Truth* illustrate that point:

> Epistemology must arrange all our beliefs, both those of which we feel convinced, and those that seem to us only more or less probable, in a certain order, beginning with those that, on reflection, appear to us credible independently of any argument in their favour, and indicating the nature of the inferences (mostly not strictly logical) by which we pass from these to derivative beliefs. Those statements about matters of fact that appear credible independently of any argument in their favour may be called "basic propositions". These are connected with certain non-verbal occurrences which may be called "experiences"; the nature of this connection is one of the fundamental questions of epistemology. . . .
>
> We said that it is the business of epistemology to arrange the propositions which constitute our knowledge in a certain logical order, in which the later propositions are accepted because of their logical relation to those that come before them. It is not necessary that the later propositions

should be logically deducible from the earlier ones; what is necessary is that the earlier ones should supply whatever grounds exist for thinking it likely that the latter ones are true. When we are considering empirical knowledge, the earliest propositions in the hierarchy, which give the grounds for all the others, are not deduced from other propositions, and yet are not mere arbitrary assumptions. They have grounds, though their grounds are not propositions, but observed occurrences.[8]

Now, we need not accept the foundationalist bias in Russell's characterization of the epistemological enterprise. What is important to recognize is that the concept of certainty as often employed by philosophers is such that propositions which are certain (Russell calls them "basic propositions" in the passages just quoted) are, ipso facto, known. Even Wittgenstein seems to acknowledge this philosophic practice when he says:

350. "I know that that's a tree" is something a philosopher might say to demonstrate to himself or to someone else that he *knows* something that is not a mathematical or logical truth.

467. I am sitting with a philosopher in the garden; he says again and again "I know that that's a tree," pointing to a tree that is near us. Someone else arrives and hears this, and I tell him: "This fellow isn't insane. We are only doing philosophy."[9]

Although Wittgenstein may have been chiding philosophers for what he would have considered pointless concept inflation, it is clear that he recognizes that they have eschewed the customary practice of attributing knowledge and certainty to distinct classes of propositions. In fact, I believe that it is reasonably clear that philosophers in general have taken a proposition to be certain when it is supported by evidence of a sort which bestows upon it the highest degree of epistemic warrant.[10]

Now, of course, some philosophers, notably Russell again, have argued that although a proposition such as 'there is a tree in my garden' can be known, it cannot be certain, precisely because the evidential grounds for it are not sufficient to confer upon it the highest degree of epistemic warrant. The reason often given is that in order for a proposition, say p, to be certain for S, it must be *logically* impossible for S to be mistaken that p on the basis of the evidence that S has for p. If we assumed that propositions expressing S's current sense experiences provided the evidential grounds for a proposition such as 'there is a tree in my garden', then, as Ayer says, "philosophers like Russell have concluded that no proposition which implied the existence of a physical object was altogether certain. What they meant was that no such proposition ever logically followed from any set of propositions which merely recorded the content of one's current sense-experiences."[11]

Similarly, Moore, while commenting on Russell's view that "material object" propositions are never certain, says:

> Russell's view that I do not know for certain that this is a pencil or that you are conscious rests, if I am right, on no less than four distinct assumptions: (1) That I don't know these things immediately; (2) That they don't follow logically from any thing or things that I do know immediately; (3) That, *if* (1) and (2) are true, my belief in or knowledge of them must be 'based on an analogical or inductive argument'; and (4) That what is so based cannot be *certain knowledge.*[12]

But as I have already mentioned, it is precisely the identification of grounds sufficient to render p certain with grounds entailing p which this chapter is designed to question. I wish to show that p can have the highest degree of epistemic warrant on the basis of some evidence, e, without e entailing p. Thus, I hope to demonstrate that Moore was correct when he said:

> I am inclined to think that what is "based on" an analogical or inductive argument, in the sense in which my knowledge or belief that this is a pencil is so, may nevertheless be certain knowledge and *not* merely more or less probable belief.[13]

Not only do I believe that propositions like 'this is a pencil' are known and absolutely certain, but I also hope to show that all the following can be certain for S:

Jones owns a Ford.

Carter was the President of the United States in 1979.

There is a brown table in my living room. (Uttered when I am *not* in my living room looking right at the table.)

Aristotle, not Plato, wrote the *De Anima.*

No human has ever eaten an adult elephant in one sitting.

In other words, I wish to show that a wide variety of empirical, contingent propositions which might be evidentially grounded by nondeductive inferences—propositions which Wittgenstein would have said were knowable but not certain—are absolutely certain.

Let me begin to clarify the notion of absolute certainty by referring to a passage in Roderick Firth's excellent article, "The Anatomy of Certainty." He characterizes what he takes to be a weak notion of certainty (a notion he attributes to Norman Malcolm) in the following:

> . . . On the basis of the distinctions already drawn, it is not difficult to define a . . . use of "certain" that would allow us to maintain that ideal irrefutability is relevant to certainty. Since to say that p is irrefutable is to imply that *not* p cannot become warranted, we can maintain that p is ideally irrefutable for S at *t* if and only if
>
> (7) There is no imaginable event such that if S were justified at *t* in believing that it will occur after *t*, *not* p would therefore become warranted for S at *t*.[14]

Richard Miller defines a similar weak notion of certainty in his recent paper, "Absolute Certainty," and claims that many empirical propositions satisfy this condition of certainty. He recognizes that a stronger characterization of certainty is definable and that a sceptical philosopher might insist upon employing it. He says:

> The third clause [a clause similar to Firth's (7)] says that reason must not, as a result of any possible new data, dictate a switch from the belief in question to its negation. It might be thought that a stronger requirement is appropriate to knowledge with absolute certainty, viz., that reason should dictate the maintenance of the belief in question in the face of any new data. Absolutely certain knowledge, it might be felt, should be the only reasonable alternative, come what may. . . . The equation of absolute certainty concerning the material world with absolutely conclusive justification surely is not obviously correct. I could not give an absolutely conclusive justification of my belief that there are automobiles. . . . If there is no strong reason for rejecting my relatively loose third condition in favour of the more stringent one, there is a reason for preferring it. If the more stringent requirement were adopted, we could not be said to know with absolute certainty that there are automobiles. . . . It seems obvious that we do know with absolute certainty that there are automobiles. A strong reason is needed for adopting a requirement on certainty that would make it slightly uncertain that there are automobiles. No such strong reason seems to be available in support of the requirement that reason dictate the maintenance of belief in the existence of automobiles in the face of all possible new data.[15]

Thus, a primary reason offered by Miller for rejecting this stronger characterization is that to adopt it would be to take the first step down the path toward scepticism. But that reason is not available to us in the present discussion. Presumably, we cannot reject a characterization of certainty simply because it purportedly supports the sceptic's position, since our task is to assess the merits of scepticism. Some philosophers—including, of course, the sceptics—would not find scepticism a consequence devoutly to be avoided.

For our purposes, Firth correctly characterizes the options which we must make available to the sceptic concerning weak and strong characterizations of certainty. Firth says:

> As soon as we spell out and classify this particular use [i.e., the analysis presented in (7)] of "certain," however, it becomes clear that it is a *weak* use of the term—weak, that is to say, in relation to other uses that are easy to define. A philosopher who is "skeptical" to the extent of maintaining that the statement "This is an ink-bottle" is never certain might nevertheless grant that the statement is sometimes ideally irrefutable—that is, that it sometimes meets the requirement defined by (7). But to be certain, he might say, it must not only be safe from *refutation* but safe from even the slightest degree of *disconfirmation.* He might maintain that p is in this sense "ideally disconfirmation-proof" for S at *t* if and only if
>
> (8) There is no imaginable event such that if S were justified at *t* in believing that it will occur after *t*, p would therefore become less warranted for S at *t*. . . .
>
> Malcolm's distinctions are illuminating, I feel, and his conclusions probably correct, when he argues that in his position no additional information would "prove" that there is no ink-bottle, or "show" that the statement is false. *But if a "skeptical" philosopher is asked what more anyone can look for than the certainty that this gives him, he is free to reply that he is looking for certainty of a kind that meets requirement (8).* [Emphasis added.] Perhaps statements like "5 × 5 = 25" and "I am in pain" are sometimes certain in this strong sense. (To discuss this would carry us far beyond the limits of the present essay). But it seems clear to me that empirical statements about the "external world" are not.[16]

Although I believe that Firth's claim that the sceptic is "free" to adopt a strong characterization of certainty is correct, there are three preliminary comments I wish to make about the characterization of certainty referred to by Firth in (7) and (8). First, I do not, of course, agree with either him or Miller that accepting this strong characterization of certainty leads to scepticism. That is one of the characterization of certainty leads to scepticism. That is one of the central points I wish to establish in Chapter Three. Second, as Firth points out, the important difference between the weak and strong notions of certainty is whether the effect of the additional information is to justify ~p, as in (7), or merely to weaken the justification of p, as in (8). There is a *seemingly* different, middle course available between (7) and (8), namely, that the additional information is such that p is no longer justified. One might think of these three notions of certainty as embodying three forms of the defeasibility analysis of knowledge. For one could characterize a justification of p as defective in any of the three following ways:

(1) the justification of p by evidence, e, is defective if and only if there is a true proposition, d, such that (d&e)C~p;

(2) the justification of p by evidence, e, is defective if and only if there is some true proposition, d, such that (d&e)₵p;

(3) the justification of p by evidence, e, is defective if and only if there is some true proposition, d, such that (d&e) weakens the confirmation of p.

I have argued elsewhere[17] for a view similar to the middle course represented by (2), and the characterization of justification developed in Chapter Two employs a notion of overriding propositions analogous to the condition of defectiveness portrayed in (2).

Now, it may be thought that because the present task is to develop an account of *absolute* certainty, the strongest condition, namely, that exemplified by (3), should be used throughout the remainder of this chapter. It is true that condition (3) seems to be more stringent than the apparently, different middle course depicted by condition (2). But I will show in section 3.12 that given the characterization of confirmation used throughout this chapter and the previous chapter, the set of propositions satisfying condition (2) is identical to the set satisfying condition (3).

There is one final important point that needs to be made about the notion of certainty contained in both (7) and (8). There is an interpretation of them—what I would take to be at least one perfectly ordinary way of understanding them—which makes both unacceptable for our purposes, because, like the Contrary Prerequisite Elimination Principle considered in Chapter Two, they lead to the requirement that the evidence for a proposition must entail it if the evidence is to be sufficient for knowledge. For consider the requirements for certainty as specified by the so-called "weak" (7). Surely there is one future *imaginable* event (if that means logically possible event) which is such that if S were justified at t in believing that it will occur after t, ~p would become justified for S at t. It is the event of S's coming to realize that there is an evil genius making S falsely believe at t that p. In fact, any *imaginable* event entailing one of the contraries of p would suffice to show that p is not certain for S. The only propositions immune to that possibility would be necessary truths. But since necessary truths are entailed by any proposition, the sceptic would be covertly requiring that, since certainty is a necessary condition of knowledge, in order for p to be known, the evidence for p must entail p.

That interpretation of (7) may not be what Firth means, since he does seem to believe that some empirical statements about the "external world" would satisfy (7) and he does once speak of "additional information" rather than "additional possible information." Also, "imaginable events" could be used to designate a subclass of (actual) events (i.e., those actual events which are imaginable or foreseeable by S). In addition, (7) could be taken to indicate a purported psychological fact about S. That is, at t, having the evidence which S has for p, S may simply be unable to imagine defeating evidence for p. Thus, the evidence need not entail p, because the limits of S's imagination rather than the degree of evidence would determine whether p is certain. (See section 3.4 for a further discussion of this psychological concept of certainty.)

But, be that as it may, the point of considering these various notions of certainty was merely to illustrate and reaffirm the task and general strategy of this chapter. To repeat: I will grant that knowledge entails absolute certainty but not that the evidence for a proposition must entail it in order for the proposition to be known and, thus, absolutely certain.

3.3 Six Desiderata (D1 Through D6) of an Adequate Account of Absolute Certainty

The previous section contained some preliminary remarks designed to clarify the general task of developing an adequate account of absolute certainty. Those remarks could be summarized as follows: Our task in this chapter is to sketch an account of absolute certainty within the confines of a defeasibility theory of knowledge and in accordance with the ground rules of the dispute between the sceptic and the nonsceptic first presented in Chapter One. If that can be done, and if the sufficient conditions of absolute certainty are such that many contingent, empirical propositions fulfill them, then we will have located a good reason for believing that S can know that p.

It is time to be more specific about what I take to be the desiderata of an adequate account of certainty. These desiderata will serve as a litmus test of my proposed partial analysis of certainty. It is a partial analysis, just as the analysis of justification presented in Chapter Two was only a partial characterization, because some of the same constituent concepts will remain relatively unanalyzed. But if the account satisfies these desiderata, I think we can assume that it provides an adequate model in which to test the claim that some empirical propositions are absolutely certain and known. The following desiderata seem to be correct:

D1 The distinction between psychological and evidential certainty must be clarified and the relationships explored.

D2 The absolute sense of "certainty" required by the sceptic should be explicated.

D3 The relative sense of "certainty" should be explicated.

D4 The relationship between the absolute and relative sense must be clarified.

D5 The entailment between knowledge and absolute certainty should be made manifest.

D6 The account must correctly capture our intuitions concerning the extension of the concept of knowledge and certainty.

The next sections are designed to clarify each of the six desiderata.

3.4 D1: The Distinction Between Psychological and Evidential Certainty

The distinction between evidential and psychological certainty should be maintained throughout the discussion. If we fail to maintain that distinction, we are liable to be inadvertently snared by some traps laid by the sceptic. We will consider some of these traps in the sections which follow.

The issue before us now is: How should we construe the distinction between psychological and evidential certainty? We can attribute beliefs to S which are more or less strongly held.[18] S can hold the belief that p with more or less tenacity. But it is crucial to note that the degree of *tenacity* is not necessarily related to the degree of *evidence* which S has for p. Some persons may *feel* certain that p on the basis of slim, inadequate, or perhaps even self-contradictory evidence. And some may *feel* certain with no evidence whatsoever. Still others may *feel* certain of anything for some alleged sceptical reasons, or merely because they are epistemically timid and never feel certain that p even though they have that epistemic right. It is this sense of 'certainty' to which Ayer was apparently alluding when he claimed that if S knows that p, then S has the right to be sure (psychologically certain) that p.[19] Psychological certainty "is simply synonymous with 'confident' or 'completely convinced'."[20]

Some philosophers have used the expression 'subjective-certainty' to refer to what I am calling psychological certainty'. For example, Wittgenstein says:

193. What does this mean: the truth of a proposition is *certain?*

194. With the word "certain" we express complete conviction, the total absence of doubt, and thereby we seek to convince other people. That is *subjective* certainty. . . .[21]

For reasons which will become obvious shortly, I prefer not to use the expression 'subjective certainty'. Instead, I will use the expression 'S is certain that p' to attribute psychological certainty to S with regard to p.

Implicit in Ayer's view mentioned above is the second sense of 'certainty'—i.e., evidential certainty. If any proposition is evidentially certain, it is the evidence for it that makes it so. I will use the expression 'p is certain' to indicate that there is conclusive evidence for p and the expression 'p is certain for S' to indicate that S has conclusive evidence for p. It is evidential certainty that is the primary concern of this chapter, since it is that sense of 'certainty' which seems most relevant to considerations surrounding scepticism.

In the conclusion of the paragraph quoted above, Wittgenstein contrasts what he calls 'subjective certainty' with 'objective certainty'.

194. . . . But when is something objectively certain? When a mistake is

not possible. But what kind of possibility is that? Mustn't mistake be *logically* excluded?

Now it may appear that Wittgenstein has granted the very point I wish to deny—namely, that absolute evidential certainty obtains only when the evidence for a proposition entails it. For he does say that something is objectively certain only when a mistake (concerning its truth) is *logically* impossible. It must be recalled, however, that Wittgenstein takes a proposition, p, to be certain (presumably in the 'objective' sense) only when there is no evidence for it because its acceptability depends upon the practices essential to the language game in which ⌜p⌝ is uttered. Thus, in the paragraphs following his introduction of the distinction between 'subjective' and 'objective' certainty he continues:

195. If I believe that I am sitting in my room when I am not, then I shall not be said to have *made a mistake*. But what is the essential difference between this case and a mistake?

196. Sure evidence is what we *accept* as sure, it is evidence that we go by in *acting* surely, acting without any doubt.

    What we call "a mistake" plays a quite special part in our language games, and so too does what we regard as certain evidence.

As I understand it, Wittgenstein's claim is that if I believe that I am sitting in my room when I am not sitting in my room, then I did not (do something as simple as) *make a mistake* like counting the same penny twice when adding up the change in my pocket or remembering the departure time of a train as being 2:15 P.M. when it is actually 2:35 P.M. or absentmindedly putting my right shoe on my left foot. If I believe falsely that I am sitting in my room, then I am either demented or the victim of some grand illusion. He considers a similar case in which it would be incorrect to attribute a mistake to someone:

67. Could we imagine a man who keeps on making mistakes where we regard a mistake as ruled out, and in fact never encounter one?

    E.g. he says he lives in such and such a place, is so and so old, comes from such and such a city, and he speaks with the same certainty (giving all the tokens of it) as I do, but he is wrong.

    But what is his relation to this error? What am I to suppose?

68. The question is: what is the logician to say here? . . .

71. If my friend were to imagine one day that he had been living for a long time past in such and such a place, etc. etc., I should not call this a *mistake*, but rather a mental disturbance, perhaps a transient one.

72. Not every false belief of this sort is a mistake.

73. But what is the difference between mistake and mental disturbance? Or what is the difference between my treating it as a mistake and my treating it as a mental disturbance?

74. Can we say: a *mistake* doesn't only have a cause, it also has a ground? I.e., roughly: when someone makes a mistake, this can be fitted into what he knows aright.

Thus, the logical impossibility of making a mistake with regard to a proposition, say p, which is objectively certain (in Wittgenstein's sense) does not depend at all on the strength of the evidence for p. To say that an utterance is 'objectively' certain is to claim that it has a particular role in the speaker's linguistic community. So, although I will be using *'psychological* certainty' in much the same way that Wittgenstein uses 'subjective certainty', my use of 'evidential certainty' is not parallel to his use of 'objective certainty'. (Since the usual contrast term with "subjective' is 'objective', I have chosen not to use the expression 'subjective certainty' in order to avoid confusing the Wittgenstein distinction with the one I will attempt to develop.)

Now, if it was correct to argue, as I did in section 3.2, that there is a perfectly recognizable sense in which philosophers have used 'certainty' to refer to those propositions, if any, whose evidence bestows upon them the highest degree of epistemic warrant, then whether propositions are 'objectively' certain is irrelevant to our inquiry. For our question is whether there are any absolutely *evidentially* certain propositions. I will continue to grant that sceptical philosophers have the right to insist that absolute *evidential* certainty is a necessary condition of knowledge.

Some philosophers, however, appear to believe that (absolute) psychological certainty is also required by knowledge, *and, in addition that such certainty never or hardly ever obtains.* This is one of those traps mentioned above which can be avoided; for I think that it will become apparent as the discussion in this chapter develops that the view that psychological certainty hardly ever obtains results from failing to distinguish carefully between psychological and evidential certainty.

3.5 D2, D3, D4: Absolute and Relative Certainty and Their Relations

What is absolute certainty? I believe that the best way to begin to answer that question is to examine Peter Unger's discussion of absolute certainty. We will see in section 3.12 that, although his discussion concerns what I have called absolute *psychological* certainty, not absolute evidential certainty, many of his comments about the former can be transferred to the latter.

Unger argues that 'certainty' is an absolute term (that is, having no degrees) like 'flat':

> I will argue that 'certain' is an absolute term, while various contrasting adjectives, 'confident,' 'doubtful,' and 'uncertain' are all relative terms. . . .
>
> With regard to being certain, there are two ideas which are important: first, the idea of *something's* being certain, where that which is certain is *not* certain *of* anything, and, second, the idea of a *being's* being certain, where that which is certain *is* certain *of* something. A paradigm context for the first idea is the context 'It is certain that it is raining' where the term 'it' has no apparent reference. . . . In contrast, a paradigm context for the second idea is this one: 'He is certain that it is raining,' where, of course, the term '*he*' purports to refer as clearly as one might like. . . .
>
> Though there are these two important sorts of context, I think that 'certain' must mean the same in both. . . . In both of our contexts we must be struck by the thought that the presence of certainty amounts to the complete absence of doubt. This thought leads me to say that 'It is certain that p' means, within the bounds of nuance, 'It is not at all doubtful that p,' or 'There is no doubt at all but that p.'[22]

Unger goes on to argue that, just as something is never (really, that is, absolutely) flat, since everything is always somewhat bumpy, nothing is (really) certain. One of his arguments for scepticism is simply that since knowledge entails absolute certainty, and nothing (or hardly anything) is ever absolutely certain, nothing (or hardly anything) is known.

Throughout his discussion, Unger correctly singles out the two important characteristics[23] of absolute terms. They can be summarized as follows:

(1)   Terms like 'flat' or 'certain' usually have relative contrastive terms like 'bumpy' and 'doubted' respectively. And it is a feature of the application of such absolute terms to an object that, if it is flat

(certain), it is *not at all* bumpy (doubted).

(2)   If something is flat (certain), then there is nothing more flat (certain).[24]

Following the strategy adopted previously, I am willing to grant that knowledge entails absolute certainty. The two general characteristics of certainty identified by Unger seem to be all that a sceptic would want to require. For they amount to requiring that if p is certain, then there are no doubts at all that p; and if p is certain, nothing can be more certain. Thus, the task is to show that these characteristics of absolute terms are captured by what I will propose as a partial account of certainty and that many contingent, empirical propositions satisfy these conditions.

In fact, I used "doubted" as the relative contrasting term for 'certain' in (1) rather than what Unger sometimes uses, i.e., "doubtful." Although an ordinary language contradictory of 'it is certain that p' is 'it is doubtful that p', the use of that contrastive expression might lead to the conflation of psychological and evidential certainty mentioned at the end of section 3.4. For the parallel with 'not at all bumpy' is not 'not at all doubtful', but rather 'not at all doubted', since the issue, *at least with regard to psychological certainty,* is not whether there are or will be *grounds* for doubting p, but rather whether S has at some given time any doubts whatsoever (whether reasonable or not) about the truth of the proposition that p. Conversations with the evil genius or other miraculous occurrences might *induce* doubts, or perhaps even imagining such events might *produce* doubts, but that is irrelevant with regard to whether S was (or is) certain that p at a time before these real or merely imagined events occurred. Something is flat if it is not at all bumpy; it is not further required that nothing could induce bumps.[25] S is absolutely certain that p, if S has no doubts whatsoever that p. It is not further required that nothing will or could induce doubts.

Richard Miller makes a similar point when he says:

> People can (undoubtedly) be made to feel sensations of doubt in connection with almost any proposition, in ways such as I have sketched. Readers of the sceptical arguments in Descartes and Hume sometimes feel surprise, bewilderment, giddiness, and vague anxiety, just as they do when they suddenly become doubtful of a cherished belief. But if I have correctly described the attitude that knowledge with absolute certainty requires,

this attitude does not preclude susceptibility to feelings of doubt. The mere possibility of producing sensations of doubt in someone in connection with a proposition will not conclusively establish that he is doubtful or less than completely certain of it.[26]

In order to clarify the concept of absolute psychological certainty, let us consider the view of Harry Frankfurt, who argues that 'certainty' is a relative term. He says:

Certainty is generally thought to be susceptible of variations in degree: people say that statements are "sufficiently certain," "more certain than ever," "quite certain," and so on. If a person regards one statement as being more certain than another he naturally thinks it reasonable to risk more on the truth of the one statement than he thinks it reasonable to risk on the truth of the other. We can imagine bets being made on the truth of various statements. It would make no sense for a person to be willing to bet one amount on a statement he regards as rather uncertain and to be willing to bet only a lesser amount on a statement he regards as certain, assuming that all bets are made at the same odds and that the person is interested only in maximizing the chances of winning his bet. Indeed a plausible way of testing whether or not a person regards one statement as more certain than another is to determine whether or not he is willing to stake more on the truth of the one than on the truth of the other. If he were not willing to do so, it would be difficult to understand what he meant by claiming that the one is more certain than the other.[27]

Ayer also notes the relative use of 'certainty' in the following:

When philosophers talk about certainty, they often treat it as a matter of degree. Thus, a philosopher who does not deny the truth of such a proposition as that there are trees in his garden, or even deny that there is a perfectly proper sense in which he can be said to know that it is true, may still wish to say that it is less certain than some proposition which records his current visual or tactual sensations.[28]

It should be apparent that both Frankfurt and Unger are speaking about what I have called 'psychological certainty' and thus are in direct disagreement. For Unger claims that 'certainty' as paradigmatically used, is an absolute term, whereas Frankfurt claims that it is, paradigmatically, a relative term.

Now, for our purposes, it is not important to determine which account of 'certainty' is correct. What is important is to recognize the distinction between absolute and relative uses of 'certainty' and to be clear about our use of it throughout the present discussion.

Just as there are absolute and relative uses of 'psychological certainty', there are absolute and relative uses of 'evidential certainty', construed as a term referring to the extent of the evidence for a belief. We often claim that our *evidence* for p makes it relatively certain that p, meaning certain enough for deciding between p and one of its contraries; but sometimes we claim that our evidence is sufficient to provide us with absolute certainty that p. It is the truth of that last claim which is usually challenged by the sceptic.

Consequently, there are four varieties of certainty which concern us:

(1)   absolute evidential certainty
(2)   absolute psychological certainty
(3)   relative evidential certainty
(4)   relative psychological certainty

As I have said, Chapter Three is primarily concerned with absolute and relative *evidential* certainty. When I granted that knowledge entailed certainty, I intended to grant that knowledge entailed absolute evidential certainty.[29] I will also grant that knowledge entails absolute psychological certainty, but since that means that S has no doubts whatsoever that p (regardless of the evidence for p), it appears to present no obstacles to knowledge. Apparent obstacles arise only when psychological and evidential certainty are conflated or when it is believed that psychological certainty leads to dogmatism. (See section 2.11.)

3.6   D5: Knowledge Entails Absolute Certainty

To satisfy D5, the entailment between knowledge and absolute certainty must be made manifest. Now, I suppose that it will be wondered why I have granted that knowledge entails absolute certainty. For although many philosophers have argued that knowledge entails certainty, perhaps an equal or greater number have disagreed; and one motivation for denying the entailment has surely been the specter of scepticism. It was believed that if knowledge entailed certainty, especially absolute evidential certainty, then Direct Scepticism was inevitable. Absolute evidential certainty, if granted at all, was reserved for knowledge of analytic propositions or the so-called "basic" (i.e., self-evident) propositions.

Thus, many philosophers would agree with Keith Lehrer's claim that scepticism is the inevitable and obvious result of granting that knowledge entails certainty. In fact, Lehrer (a onetime sceptic—see Chapter Two, especially section 2.1) says:

> The sceptic is correct, we concede, in affirming the chance of error is always genuine. . . . To sustain scepticism, a sceptic must go on to argue that if there is some chance that S is incorrect in his belief that p, then S does not know that p. On the analysis of knowledge we have articulated, this premiss is unavailable. . . . He may thus know that p, and know that he knows, even though there is some chance that p is false, and S believes this to be so. Hence we may accept the premiss of the sceptic concerning conceptual change and the universal chance of error implicit therein without accepting the deep sceptical conclusion of universal ignorance. . . . Thus our theory of knowledge is a theory of knowledge without certainty. We agree with the sceptic that if a man claims to know for certain, he does not know whereof he speaks.[30]

Now, I do not intend to rehearse the arguments for the claim that knowledge entails certainty. I am prepared to grant that entailment because by so doing, and by showing that an acceptable explication of certainty is inherent in a defeasibility account of knowledge, the case against scepticism will be strengthened. In addition, if there are no valid general objections to the defeasibility theory, the sceptic ought to be willing to accept it as a proper analysis of knowledge. As stated before, the issue then becomes: Are the necessary and sufficient conditions of knowledge sometimes fulfilled?

To use Lehrer's terminology, I am willing to grant that if S knows that p, then although the justification is defeasible, there is no *real* chance of it being defeated. Modal ambiguities are plentiful here, and I hope to show that the sceptic has conflated the logical possibility of error with the real chance of being wrong. Put another way, I hope to show that some of our justifications are such that they do *in fact* guarantee our beliefs in the actual world, although they do not guarantee those beliefs in all *possible* worlds. As we will see, the latter would again require that the evidence for our belief entails them.

3.7 D6: Intuitions About Certainty

Pretheoretical intuitions are not completely trustworthy guides for determining the acceptability of a philosophical account, if only because it can be doubted that any one of us (philosophers) has such intuitions. As I pointed out in section 3.2 with regard to the concept of certainty, philosophers customarily take concepts from their ordinary context and imbue them with new meanings. Some decry that practice in general, believing that philosophers who engage in it are just "spinning wheels" which are not connected to anything substantial. Others think that this customary practice creates no insurmountable problems provided that the philosophers' methodology is sufficiently self-conscious. It should be obvious that my allegiance lies with the latter.

This is not the proper place to embark on a general defense of that practice. But I think it is safe to assume that anyone with steadfast contrary loyalties will have put this book aside well before reaching this point!

As mentioned above, we must acknowledge that wholly pretheoretical intuitions are rarely, if ever, available to us. In particular, since the concepts which we have been exploring (justification, knowledge, certainty) have been endowed with new or at least vastly extended uses, it is not at all clear that there are any relevant, well-defined, pretheoretical intuitions concerning them.

Nevertheless, it does seem clear that philosophers employ what might be called a set of paradigmatic cases which incorporate widely shared presuppositions against which philosophic accounts can be measured. For example, in ethics it is assumed that it is worse to harm one hundred people than one person; in metaphysics it is presupposed that an object can change and nevertheless remain the same in some sense; and in epistemology it is generally believed that S knows that p only if p is true. If an account is acceptable on the basis of all other philosophical standards (e.g., coherence, simplicity, generalizability), then if, in addition, it accords with these generally accepted views, it gains more credibility.

Of course, it should go without saying that often the paradigmatic cases are reinterpreted or the generally accepted views revised in the light of a new philosophic account. Almost no philosophic position, however extraordinary, has remained unchampioned. For example, one can think of philosophers who have claimed that the widely accepted views mentioned in the preceding paragraphs ought to be revised in light of a philosophic theory. But if such a theory is to succeed in reshaping these generally accepted views, it must present compelling grounds. It might, for example, uncover certain questionable, although hitherto unquestioned, presuppositions concerning these views or it might discover some inconsistencies in our generally accepted notions about the paradigm cases.

Thus, I will take it that the analysis of knowledge and certainty developed in this chapter should either accord with our views concerning the relevant paradigm cases, where such views can be identified, or it should demonstrate why those views should be revised. Various cases are examined throughout the remainder of this chapter, but I think it would be useful to mention a few specific discussions at this point:

(1) In section 3.8 and 3.9, I will show how my account accords with the consensus concerning a variety of cases present in the literature about the extension of knowledge: the Grabit Cases, the Civil Rights Worker Case, the Clever Car Thief Case, and the Gettier Cases.

(2) In sections 3.12, 3.13, and 3.16, I will try to persuade the reader that some revisions are required in the consensus concerning the extension of propositions held to be absolutely certain. Specifically, I will argue that propositions such as 'Jones owns a Ford' are no less certain than 'I have a hand' or '17+18=35'.

(3) In section 3.15, various forms of the Lottery Paradox will be examined. Although there may be no consensus concerning those cases, I will argue that the results of applying my analysis of justification, knowledge, and certainty are desirable.

3.8 The Defeasibility Theory of Knowledge and Absolute Certainty

The six desiderata of an adequate account of certainty should now be relatively clear. I hope they will become still more clear as the discussion in this chapter proceeds. What remains to be done is "merely" to show that the defeasibility theory of knowledge provides a basis for an account of certainty which satisfies them!

I have argued elsewhere for a particular formulation of a defeasibility analysis of knowledge which I claimed was superior to the other defeasibility accounts and to the causal theories as well.[31] I would like to develop that particular formulation of the defeasibility theory still further in light of the model of justification presented in Chapter Two. For I believe that this formulation of the theory has one further virtue, namely, it provides a basis for an explication of certainty within the confines of the ground rules delineated in Chapter One. Specifically, the task is to show that knowledge that p, as portrayed by the defeasibility theory developed here, requires that p be absolutely certain on the basis of some evidence, e, but that e need not entail p. Along the way, I will continue to grant to the sceptic whatever can reasonably be granted. Later in this chapter, I will argue that many empirical, contingent propositions are known and, a fortiori, absolutely certain.

The role of this section is similar to that of section 2.7 in Chapter Two, in which the model of justification was developed. That is, I hope that the discussion is interesting in its own right, but its usefulness here is primarily to test my proposed account of certainty against the six desiderata.

Before presenting and defending that particular model of the defeasibility theory, perhaps a few comments are appropriate about the defeasibility theory in general and its relationship to the sceptic's demands. For although the account of certainty will be developed in light of a particular analysis of nondefective justification, it is the common features of the defeasibility theories of knowledge which are primarily responsible for their ability to portray accurately the conditions in which a proposition is absolutely certain.

The defeasibility theory, like scepticism, maintains that it is necessary for a belief to be true and justified in order for it to be certifiable as knowledge. And also like scepticism, it holds that true, justified belief is not sufficient for knowledge. For even though S may have such a belief, there may also be information unknown to S which makes S's justification defective. In the clearest of those cases, the new information may be such that it shows that S arrived at the truth by an epistemologically lucky accident. And if we demand that the evidence, e, render p absolutely certain, there may be information which S does not possess which is such that were it conjoined with e, the conjunction, d&e, would no longer confirm p.

Let us examine a clear case first. Suppose that S looks at the gas gauge of her car and that it reads "¼." As a result, S comes to believe that the tank is one-quarter full. Suppose further that the tank is, in fact, one-quarter full and that the fuel guage has been accurate in the past. Thus we may assume for the moment that S has a true, justified belief that the tank is one-quarter full. That is, let us assume that p is true and satisfies the conditions of justification presented in Chapter Two, especially section 2.7.

Now, before using this case to clarify the conditions under which a justification becomes defective, it would be useful to pause briefly to review some of the more important features of that model of justification, since much of what will be said concerning defective justification will rely upon it. We said that a proposition, say x, is justified for S if and only if there is a proposition, say w, available for S to use as confirming evidence such that w confirms x (wCx); and there does not exist a proposition, say u, available for S to use as an overrider of the confirmation of x. The proposition, u, would override the confirmation of x, if, when conjoined with an evidential ancestor of x, say y_i, it was such that $(u \& y_i) C y_{i+1}$. We called a proposition available for S to use as confirming evidence a *grounded* proposition for S. All grounded propositions were also available for S to use as overriders, but there were propositions available for S to use *only* as overriders. We call such propositions *psuedogrounded* propositions for S. The conditions of groundedness were made as stringent as possible, and those for pseudogroundedness as relaxed as possible, in order to accommodate the sceptic's demands. For example, a grounded proposition had to be obtained by S in an epistemically reliable manner or located on a nondegenerate chain of propositions each link of which was confirmed by the preceding link and which was anchored in a subset of S's reliably obtained beliefs which we called the Γ_S^R set. But a proposition need not be reliably obtained or anchored in Γ_S^R by a nondegenerate chain in order for it to be available as an overrider for S. It could simply be any member of the set of beliefs actually subscribed to by S, what we called the B_S set, or linked by a nondegenerate chain to a variety of subsets of B_S. (See section 2.7 for more details.) In addition, we allowed overriders, but not confirmers, to be "manufactured" by conjoining any grounded or pseudogrounded proposition, as long as the resulting chain did not degenerate.

Finally, although the concept of confirmation used in the characterization of justification was not completely clarified, primarily because the differences in various plausible models were not decisive in the dispute between the sceptic and the nonsceptic, some restrictions were placed upon it. For our purposes here, the following are the most important:

(1) Although it is possible for xCy and for $x \rightarrow y$, it is not required that $x \rightarrow y$ in order for xCy.

(2) Confirmation, like justification, is reflexive, nonsymmetrical, and nontransitive.

(3) For every x and y, $(x \& \sim x)Cy$ [even though $(x \& \sim x) \rightarrow y$].

(4) For every x and y, $(x \& \sim y)Cy$ [even if y is necessarily true or $(x \& \sim y)$ is necessarily false].

I will have more to say about confirmation in this chapter, especially in sections 3.12 and 3.15.

Now, let us return to the Gas Gauge Case by employing this model of justification. We stipulated that the proposition, let us call it h_1, *the tank is one-quarter full,* is justified for S on the basis of the proposition, call it e_n, *the normally reliable gauge reads "¼."* Thus, e_n is grounded for S, and there is no other grounded or pseudogrounded proposition for S, say u_1, such that u_1, conjoined with any proposition, say e_i, in the evidential ancestry of e_n results in it being the case that $(u_1 \& e_i)Ce_{i+1}$.

But any defeasibility theory of knowledge would have us note that, in addition to stipulating that p is true and justified for S, we could also make the justification defective by assuming that the gauge, though normally reliable, is not working properly on this occasion. In fact, let us suppose that it is stuck on the "¼" mark. Call the proposition expressing that fact d_1. Thus the defeasibility theory of knowledge would hold that, although S's belief that h_1 is true and justified, it is not knowledge. The reason, *roughly* put, is that there is additional information which is neither grounded nor pseudogrounded for S which defeats S's justification. That is, S's justification is defective because there is some other true proposition, d_1 in this case, which when conjoined with the original evidence is no longer adequate to confirm h_1. To use the terminology we have adopted, d_1 is a defeater of the justification of h_1 because d_1 is true and $(e_n \& d_1)$ Ch_1. Defeaters function analogously to overriders, but they are neither grounded nor pseudogrounded for S.

We will see shortly that this rough way of characterizing defeating propositions must be revised in a variety of ways. But the underlying intuition that knowledge entails true, nondefectively justified belief remains as the foundation of all defeasibility theories of knowledge and is, as we shall see, primarily responsible for their ability to provide the basis for an analysis of evidential certainty. For what I wish to show in this chapter is that the requirement that a justification be nondefective is equivalent to the requirement that a justification be strong enough to provide absolute evidential certainty. This is one of the main themes of Chapter Three, i.e., that nondefective justification and certainty are equivalent.

That there is at least some promise that those two concepts are equivalent should already be obvious. For the account of knowledge proposed by the defeasibility theory does not require that the adequate confirming evidence, e, for a proposition, p, entail that proposition in order for it to provide S with a justification for p. But at the same time, the defeasibility theory requires that e be related to the *total* set of information in some specified way so as to make the justification absolutely attack-proof, i.e., absolutely nondefective. The promise that the concepts of nondefective justification and evidential certainty are equivalent can be fulfilled if the characterization of nondefective justification satisfies the six desiderata of an adequate account of certainty. Consequently, the issue now becomes: How should we make precise the characterization of nondefectiveness? That is, how must the evidence, e, for a proposition be related to the total set of information if e is to provide S with a nondefective justification?

Defeasibility theorists differ in the way in which they characterize that relationship. Results obtained in Chapter Two can be employed to show that some of these accounts are unacceptable. For example, during the discussion of Dretske's arguments against the Contrary Consequence Elimination Principle, we discovered that we cannot require that e confirms p only if e (alone) confirms the negation of every contrary of p or confirms the negation of every defeater of S's justification for p. Recall that, in the Zebra Case, although I argued that whenever S is justified in believing that p, S is justified in believing the negation of every *contrary* of p (but not necessarily on the basis of e), I also argued that the *Defeater* Consequence Elimination Principle was false. S may be justified in believing that p without being justified in believing the denial of every defeater of the justification for p. In addition, in section 3.2 I argued that, for reasons similar to those which led to the rejection of the Contrary Prerequisite Elimination Principle, we cannot require that e be related to p in such a way that no *possible* evidence could defeat the justification of e by p. To do so would lead to the unwanted consequence that e must entail p.

So how should we characterize the relationship between grounded and pseudogrounded propositions for S and the information which S does not possess which makes the justification defective? It will be useful to recall here the *rough* characterization of certainty suggested in section 3.2: A proposition, p, is certain on the basis of e if and only if there is no new information such that the conjunction of it with e fails to justify p. Note the obvious parallel between the rough characterization of certainty and the rough characterization of non-defectiveness. We should continue to exploit that parallelism in order to show that the two concepts are equivalent.

We have just seen that, in the Gas Gauge Case, there may be other true propositions (i.e., *actual* counterevidence), neither grounded nor pseudogrounded for S, which are such that, even though S arrived at a true belief, p, it can be seen as a felicitous coincidence. That is, S's evidence, e, for p may justify S in believing that p, but, when the total set of information is viewed, it is merely a lucky accident that S arrived at the truth. There is other information which, had S gained along with e or gained before e, S would not be justified in believing that p. S has fortuitously acquired a subset of the total set of information which, although leading S to a true belief (namely p), nevertheless is insufficient to certify S's belief as knowledge.

Consider another example. S may believe that she owes $500 in federal income taxes. Suppose that it is true that she owes $500; and that S was careful enough in following the procedure on the forms and performing the mathematical computations to justify the proposition that she owes $500. Now, even though S was sufficiently careful, she could have made an error, or even several errors. According to the defeasibility theory of knowledge, if S had made two errors—one canceling the other—S would not have known that she owed $500; for having the correct, justified belief was a felicitous coincidence. S does not know that she made some calculation errors, but if S were to come to realize that several mistakes had been made, S would lose the justification for the proposition that she owes $500.

As mentioned earlier, it may be thought that we can accept the rough characterization of certainty and simply require that there be *no* such defeating counterevidence, i.e., that there be no information such that, when conjoined with S's evidence, it fails to confirm the once-justified proposition. Thus, it may be thought that a justification of p by e is defective just in case there is some true proposition, d, such that $(d \& e)\mathcal{C}p$.

Although that is roughly correct, it will not quite do, because, as we shall see, it is both too weak and too strong. Consequently, several adjustments must be made. Some of them are designed to accommodate sceptical demands; others are necessary in order to bring the analysis in line with the intuitions mentioned in section 3.7. The discussion concerning these adjustments is unavoidably complex, but since the defeasibility condition plays a crucial role in the ability of my proposal to meet the six desiderata, we must proceed carefully.

Let us begin amending the rough characterization by considering what makes it too weak. Suppose that S is justified in believing that h, on the basis of some evidence, e_n. Now, by our model of justification there must be a nondegenerate chain anchored in Γ_S^R terminating in h_1 with e_n being the link immediately preceding h_1. Each link is confirmed by the preceding link. The chain looks like this:

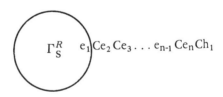

Figure 4.

According to the rough characterization, if the justification of h_1 by e_n is defective, there must be a true proposition, d_1, such that $(d_1 \& e_n)\mathcal{C}h_1$. But a weak or defective link may occur in the chain prior to e_n. Presumably the justification would be just as defective if d_1 were such that it broke the chain at any link prior to h_1. Just as we had to examine the evidential ancestry of a proposition in order to determine whether its confirmation was overriden, we must inspect the entire chain in order to determine whether it is defective. Thus, we will have to amend the condition of defectiveness by counting d_1 as a defeater of the justification of h_1 by e_n if there is some e_j such that $(d_1 \& e_j)\mathcal{C}e_{j+1}$.

This suggests another way in which even this strengthened condition remains too weak. For suppose that d_1 does not combine directly with an evidential ancestor of h_1, but rather, in conjunction with some grounded or pseudogrounded proposition for S, say g_1, it is able to break the chain at some point. Let us call the set of grounded or pseudogrounded propositions for S the E_s set. Thus, $g_1 \epsilon E_s$. Our supposition is this: Assume that there is some member of E_s, say g_1, such that $(g_1 \& d_1)$ conjoined with some e_1 fails to confirm e_{i+1}. The sceptic, no doubt, will insist that, although g_1 may not have played a role in the justification of h_1, it can be employed along with d_1 to defeat the justification. The sceptic will claim that S's set of grounded and pseudogrounded propositions, the E_s set, is defective with regard to the justification of h_1. Our policy of granting whatever can reasonably be granted to the sceptic requires that we strengthen the condition to allow for this way in which a justification of h_1 by e_n may be defective. Thus, we must allow d_1 to be a defeater of h_1 by e_n if $(d_1 \& g_1 \& e_j)\mathcal{C}e_{j+1}$. In this case we can call d_1 the *initiating* defeater and $(g_1 \& d_1)$ the *effective* defeater. The proposition, d_1, is the new information which is initially responsible for defeating the justification, but it must work through a member of E_s in order to be effective.[32]

Once having recognized the distinction between initiating and effective defeaters, however, it is obvious that effective defeaters can be produced in a variety of ways from initiating defeaters. We should pause to consider two of the more interesting ones.

Recall the case just considered. Instead of adding the proposition 'S made an error' to E_s, suppose that we add a true proposition that a CPA (Certified Public Accountant) sincerely asserted that S had made a serious mistake. Let us call this case *the CPA Case*. That proposition at least renders plausible some further proposition, say d_2, namely, *the CPA believes that S did make a mistake*, which in turn renders plausible d_3, *S did make a mistake*, which is such that $(d_3 \& e_n)\mathcal{C}h_1$. (The reason for employing the weaker epistemic relation "renders plausible" rather than "confirms" will be discussed shortly.) The sceptic will again insist that in the CPA Case, the true proposition *the CPA asserts that S made a serious mistake* is an initiating defeater and that d_3 is an effective defeater. Thus, one way in which effective defeaters may be produced by initiating defeaters is through non-degenerate chains (linked by the weaker 'renders plausible') beginning with d_1 and ending with d_n where d_n is such that $(d_n \& e_i) \mathcal{C} e_{i+1}$.

The reason for employing the relatively weak epistemic predicate, 'renders plausible' (let us call it "R"), in the chain from d_1 to d_n is that it is necessary in order to capture the distinction between initiating and effective defeaters in a way which will satisfy the sceptic. Let us define the weaker relation this way: A proposition, p, is *rendered plausible* by e if and only if e makes p epistemically more reasonable to believe than ~p.[33] Note that p may be rendered plausible by another proposition without p being plausible for S on the basis of the totality of S's evidence. Thus, an effective defeater can render one of S's justifications defective, even though the effective defeater is not plausible for S on the basis of all of S's evidence. The qualification "epistemically" is added to "more reasonable" for the reasons mentioned in section 2.16.

The predicate 'R' is weaker than the predicate 'C' because if eCp then eRp; but the converse is not true. In the CPA Case just discussed, the proposition (d_2), expressing the fact that the CPA believes that S made an error, may not confirm the proposition (d_3) that S did make an error, but it does render it plausible. In addition, the plausibility of d_3 on the basis of d_2 seems sufficient to defeat the justification of h_1. At least, the sceptic will think so, and we are obliged to accept that reasonable demand.

The second interesting way in which an initiating defeater, d_1, may lead to an effective defeater, d_n, is that d_1 may combine with some proposition in E_S, say g_j, to render plausible a proposition, d_2, and d_2 may render plausible another proposition, d_3, or d_2 in conjunction with some member of E_S, say g_j, may render plausible d_3 . . . (etc.) That chain may terminate in a proposition, d_n, such that $(d_n \& e_i) \mathcal{C} e_{i+1}$).

There will be various combinations of these two interesting ways in which an initiating defeater may lead to an effective defeater through a chain of propositions. Let us call such chains *defeater chains*, or *D-chains*, in order to distinguish them from chains linked by confirmation. (Let me add parenthetically that the sceptic has the right to insist that a similar modification be required in the characterization of overriding propositions available for S. That is, he/she may claim that if a proposition is attached to a member of the set of propositions permitted to anchor overriding chains by a non-degenerate chain of propositions each of which is rendered plausible, then it is available as an overrider for S. I did not discuss that suggestion in section 2.7 because it was irrelevant to the issues at stake there. For that modification could be made without affecting the argument in that section.) We can say that d_1 is an *initiating defeater* of the justification of h_1 by e_n for S if and only if (1) d_1 is true and not a member of E_S, and (2) there is a D-chain from d_1 to an effective defeater, d_n, of the justification of h_1 by e_n for S. The proposition, d_n, is an *effective defeater* of the justification of h_1 by e_n for S if and only if $(d_n \& e_i) \mathbb{C} e_{i+1}$. And, finally, we can say that a chain of propositions is a *D-chain* if and only if it is a nondegenerate chain such that each link, say l, is rendered plausible by a preceding link or l is a conjunction of propositions each conjunct of which is either in E_S or a proposition rendered plausible by some preceding link.

One crucial point to note here is that the initiating defeater may act indirectly by rendering plausible an effective defeater through the use of some member of E_S or by rendering plausible an effective defeater without employing any proposition in E_S. Of course, the initiating defeater may act directly to break the chain from e_1 to h_1, as in the Gas Gauge Case. That possibility is encompassed by our definition, because it allows the initiating defeater and the effective defeater to be identical. For the R-relation, like confirmation and justification, is reflexive.

A second important point to note is that, although the initiating defeater must be true, because it is "new information" not yet possessed by S, the effective defeater need *not* be true. The effective defeater could, for example, be a conjunction composed of some false member of E_S and the (true) initiating defeater. More generally, it may be a false proposition rendered plausible by some true, preceding link in the D-chain.

These features of the distinction between initiating and effective defeaters have an important consequence. For they show that our characterization of defective justification is now too strong! For only some false propositions connected by D-chains to new information can be allowed to serve as links in the D-chain terminating in an effective defeater.[34] To see this, reconsider the CPA Case, but this time suppose that S realizes that she may make a mistake in computing the taxes owed and, as a result, asks a CPA to check the return. Suppose that S's name is 'Loretta'. The CPA examines the return and finds no errors; but in writing to Loretta the CPA inadvertently leaves out the "not" in the note. Thus, there is a note which reads, "Loretta, I have checked your return carefully and it does contain errors." Now the proposition, d_1, *the CPA wrote a note which said, "Loretta . . . errors"* is true and not a member of E_S. It would render plausible the proposition that the CPA believes that S made an error, and that, in turn, renders plausible a proposition, d_n, which is an effective defeater of the justification of the proposition that S owes \$500. The proposition, d_n, is *S made an error in computing her taxes.* Thus, d_1 appears to be an initiating defeater because it is true, not a member of E_S, and linked by a D-chain to an effective defeater, d_n. But in this case—call it the *Slip-of-the-Pen CPA Case*—it hardly seems as though S's justification is defective.

It is important to recognize that the issue is not whether S would continue to know that she owed \$500 *after* actually reading the note from the CPA. For, presumably, S may cease to believe that she owes \$500 or may cease to be justified in believing that proposition if d_n becomes a member of E_S or if we amend the characterization of overriding chains as suggested parenthetically above. But whether a justification is defective does not depend upon S's actually acquiring the new information. The issue is only whether the conjunction of the new information with E_S is such that an effective defeater is produced. Thus, the proposition that the CPA wrote a note which reads "Loretta . . . errors" is an initiating defeater even were S never to see the note—because, for example, the CPA caught the slip of the pen in time and destroyed the note.

Nevertheless, as mentioned above, it seems to be an extremely counterintuitive result that S loses knowledge because of the slip of the CPA's pen. As has been pointed out by Bredo Johnson, if a normally epistemically reliable person wished to deprive any S of knowledge that h, all that would be required would be that he/she deny some important evidence in the chain leading from Γ_S^R to h.[35]

Thus we must have a way of distinguishing between misleading initiating defeaters and genuine initiating defeaters. We can begin by noting that, in the Slip-of-the-Pen CPA Case, it is not really the addition of *new information* to E_S which defeats, but rather the addition to E_S of the *misinformation* that the CPA believes that S made an error which causes the mischief. That, in turn, renders plausible the false effective defeater that S made a mistake in preparing her taxes. It must be remembered, however, that we cannot simply call every initiating defeater a misleading one if it leads to a false effective defeater, because the conjunction of the initiating defeater and a false member of E_S may result in a false effective defeater. In that case, we have granted the right of the sceptic to insist that S's justification is really defective. In addition, there are plausible variations of the CPA Case in which the initiating defeater, d_1, renders plausible some false proposition, say d_2, without employing any proposition in E_S, and d_2, in turn, renders plausible a true effective defeater. It would seem that d_1 would still be a misleading initiating defeater in that case. Hence, there may be initiating defeaters which are misleading even though the D-chain terminates in a true effective defeater. Of course, in that case there is a genuine initiating defeater—namely, the true effective defeater itself.

But there is a common feature in all of the cases in which the initiating defeater is misleading. For in each case, the initiating defeater renders plausible a false proposition in the D-chain before employing any false member of E_S. We have to say "false member of E_S" because it is at least plausible to assert that in the CPA Case the proposition that letters generally express beliefs and the proposition that S's name is "Loretta" are implicitly employed by S from her E set in order to render plausible the proposition that the CPA believes that she made a mistake. Those are, of course, true members of E_S. In other words, *misinformation which does not depend upon any false proposition in* E_S is an essential link in the D-chain leading to the effective defeater in the cases in which the initiating defeater is misleading. The defeating effect of the initiating defeater is essentially parasitic upon misinformation which does not depend upon any false proposition in E_S.

Let us suppose, then, that an initiating defeater, d_1, is *misleading* if and only if there is some false proposition, f, in *every* D-chain between d_1 and an effective defeater, d_n, and f occurs in a link in the D-chain prior to every link in which a false proposition in E occurs. All initiating defeaters that are not misleading are *genuine* initiating defeaters. We can now say that a justification of h by e for S is *defective* if an only if there is a genuine initiating defeater of the justification of h by e for S.

This characterization of the distinction between genuine and misleading initiating defeaters will be tested shortly as part of the overall assessment of my proposed analysis of knowledge. And we are *almost* in a position to consider that proposed analysis.

In order to present that analysis soon, let me assert without argument that S knows that p only if S's belief that p is true.[36] This has been questioned by some philosophers, who have argued, among other things, that the utterance ⌜S believes that p⌝ and the utterance ⌜S knows that p⌝ are such that occasions when one is appropriate are not occasions when the other is appropriate. What I said in section 3.2 about the relationships between certainty and knowledge which interest philosophers can be transferred here. In addition, the sceptic will insist upon the greatest number of plausible necessary conditions of knowledge in order to reduce the likelihood that knowledge is obtainable. That knowledge entails true belief seems to be among those demands which we should grant.

There is one point, however, which must be mentioned before presenting the proposed model of knowledge. Recall that we have distinguished two types of absolute certainty—absolute evidential and absolute psychological certainty. When I said, earlier in this section, that absolute certainty and nondefective justification were equivalent, I was referring to evidential certainty. I have stated in another section (see section 3.5) that I was willing to grant that S had to be absolutely (psychologically) certain that p in order to know that p. Thus, we may suppose that S knows that p only if S is absolutely (psychologically) certain that p and p is absolutely (evidentially) certain for S. But since a person may be absolutely (psychologically) certain that p on the basis of evidence which does not make p (evidentially) certain for S, even though S has evidence which makes p evidentially certain, the sceptic will correctly point out that both conditions may be fulfilled but that S will still lack knowledge. For S actually could have arrived at the belief that p through a process which is not absolutely attack-proof. That is, S could have arrived at p through an epistemically felicitous coincidence. Consequently, we must require that S's belief that p results from actually having employed a nondefective justification for p. To use Ayer's terminology mentioned earlier, S has the right to be sure that p only when S's belief that p is brought about by having used an epistemically satisfactory process.

Here, then, is my suggestion for the model of knowledge:

S *knows that p* if and only if

  K1 p is true.

  K2 S is certain that p on the basis of some proposition, e.

  K3 e justifies p for S.

  K4 Every initiating defeater of the justification of p by e for S is a misleading initiating defeater.

Before concluding this section, let me point out that this is a general model of knowledge designed to include both inferential and noninferential knowledge. That it includes the former is obvious; but that it also applies to noninferential knowledge may not be quite so apparent.[37] Nevertheless, once it is recalled that justification and confirmation are reflexive and that only propositions which are grounded for S can be confirmers, it should be clear that, if there are any propositions which are known without inferences from other propositions, they will be in Γ_S^R. For those are the propositions obtained by S in a reliable manner which do not depend upon other propositions for their reliability unless those other propositions depend reciprocally on them (see section 2.7). Consequently, if there is noninferential knowledge, e and p would be identical.

I will focus on the ability of the model to provide an adequate analysis of inferential knowledge, since it is primarily that which is questioned by the sceptic. Specifically, my claim is that this characterization of inferential knowledge is, in general, correct and that it is able to provide a basis for an analysis of certainty acceptable to the sceptics. Roughly put, I will argue that if K3 and K4 are satisfied, p is *absolutely evidentially* certain. In other words, if a belief that p is nondefectively justified, then p is absolutely evidentially certain. In addition, since the strength of the belief portrayed in K2 is such that S has no doubts whatsoever that p, it captures the requirements for *absolute psychological* certainty.

But in order to assess the success of this model of the defeasibility theory in providing a basis for the analysis of certainty, especially evidential certainty, more needs to be said in defense of K3 and K4. That is the task of the next section.

3.9 Further Discussion of Defective Justification and a Defense of the Characterization of Misleading and Genuine Initiating Defeaters

In this section, we will consider one alternative method of solving the difficulties raised by misleading initiating defeaters. I will argue that if the "alternative" strategy is acceptable, it is only because it implicitly employs the solution developed in the previous section. In addition, I will defend my characterization of the distinction between misleading and genuine initiating defeaters against several plausible objections. Thus, the purpose of this section is to clarify my proposal by contrasting it with one "alternative" and to demonstrate that it correctly characterizes a set of paradigmatic cases—including some which indicate that the concept of knowledge is "vague." In other words, the argument in this section is an important part of the attempt to show that my analysis of justification, knowledge, and certainty satisfies the sixth desideratum (see section 3.7).

Let us begin with the purported alternative approach to the problem of misleading information which defeats a justification but does not render it defective. It was developed by John Barker in his paper "What You Don't Know Won't Hurt You?"[38] In order to understand Barker's approach, let us for the moment ignore the distinction between initiating and effective defeaters and the variety of ways in which an initiating defeater can combine with members of E_S in order to produce links of the D-chains. Instead, let us simply consider a defeater, say d, of the justification of p by e to be a true proposition not already part of S's belief set such that $(d\&e)\mathcal{C}p$. In other words, let us begin with what I called in the previous section the "rough characterization" of a defeater.

For reasons similar to those discussed in connection with the Slip-of-the-Pen CPA Case, Barker believes that it is necessary to distinguish two types of defeaters—those which really render a justification defective and those which only appear to do so. He agrees that there are some defeaters which do not destroy a justification. For example, the proposition that there is a note written by the CPA which reads, "Loretta, I have checked your return carefully and it does contain errors" is such a proposition when it is also true that the CPA does not really believe that there are any errors and inadvertently omitted the "not" in the note.

Roughly put, Barker's suggestion is that if the *original* justification of the proposition, although defeated, can be restored by another proposition, the justification is not defective. As he says, S's true belief that p is knowledge "if and only if there is some way that any other true proposition besides p could come to be justifiably believed without destruction of the original justification for believing p."[39] Only if the justification were destroyed beyond restoration would it really be defective. On the other hand, defeated but restorable justifications are sufficient for producing knowledge.

In the Slip-of-the-Pen CPA case, a way to restore the original justification is to combine the explanation of the slip of the pen with the defeater. However, if we suppose that S really did make some errors in computing the tax, even though, as luck would have it, she arrived at the correct amount, the conjunction of the proposition that the errors were mutually canceling with the defeater that some errors were made does not restore the *original* justification; a new justification is produced for the claim that S owes $500. In short, defeated-destroyed justifications are defective, and defeated-restorable justifications are not defective.

An additional merit claimed by Barker for his proposal is that it captures the inherent vagueness of the extension of the concept of knowledge and correctly portrays the conflicting intuitions which arise in the vague cases. I believe that he is correct about the existence of vague cases, and the sixth desideratum requires that any proposed analysis of knowledge must be able to explicate these cases correctly. Thus, we will have to compare the success of Barker's proposal and my suggestion developed in the previous section for handling these vague cases.

In order to determine whether this alternative proposal is acceptable, some questions need to be asked about the distinction between defeated-destroyed justification and defeated-restorable justification. The fundamental issue is this: What are the general conditions in which a proposition restores the *original*-but-defeated justification, and when does the proposition provide for a *new* justification? Without a clear way of answering that question, the success of Barker's proposal remains in doubt. For there will be no way of determining when a justification is really defective. The proposal developed in the previous section does have a systematic method for distinguishing genuine and misleading initiating defeaters. Only misleading ones are such that every D-chain leading to an effective defeater has a false proposition in a link prior to every link in which a false proposition from E_s occurs. I hope to show that the required distinction between defeated-destroyed justification and defeated-restorable justification can be explicated only in light of the distinction between misleading and genuine initiating defeaters.

Let me turn to some specific cases which Barker uses to illustrate his proposal. In what he calls the "first version" of the Tom Grabit Case—a case in which Tom's demented mother falsely asserts that he has a twin—he claims that the *original* justification *is restored* when we discover that her demented condition was responsible for the fabrication. In the second version—the version in which there really is a twin—he claims that the *original* justification is *not* restored when we discover that the twin was in some location other than the particular place where the Tom-like-looking person was observed stealing the book. Accordingly, in the first case, there is a restorer of the *original* justification, and S has knowledge; in the second case, there is no restorer of the original justification, and S fails to know. We will return to these versions of the Grabit Case shortly. To repeat, the question before us now is: How are we to determine *in general* when a proposition merely restores the original justification and when it provides for a new one?

Let us look at two clear cases in which an original justification would be destroyed beyond restoration. Consider a case in which:

(1) S is justified in believing that p on the basis of e.

(2) There is a true proposition, say d, such that $(d \& e) \not\mathbb{C} p$.

(3) There is also a true proposition, say n, and $(d \& e \& n) C p$, but n, alone, confirms p, i.e., nCp.

This is a clear case in which S does not know that p on the basis of e. For it is apparent that n does not restore the *original* justification for p—it provides for a completely new one. The evidence, e, is not used at all in the conjunction, $(d \& e \& n)$. In our terminology, d would be a genuine initiating (and effective) defeater. This is a schematization of one of the original Gettier cases.[40] The Gas Gauge Case discussed in the previous section is also such a case if we let 'e' be the proposition that the normally reliable gauge reads "¼" and let 'p' be the proposition that the tank is one-quarter full. Now if 'd' is the proposition that the gauge is stuck on the "¼" mark, and 'n' is that a new gauge was installed and it reads "¼," it should be clear that the original justification is not restored by n, but rather that n provides S with a completely new justification.

There are other cases in which the original evidence, e, is a conjunction, say $(e_1 \& e_2)$, in which the first conjunct is true and the second one is false. Let us suppose that e_1 (the true conjunct) is insufficient by itself to justify p. The proposition $\sim e_2$ would be an initiating (and effective) defeater. Now, suppose that n conjoined with e_1 is sufficient to confirm p. Even though S would now be justified in believing that p, it is clear that the *original* justification has not been restored. A part of the original justification, namely e_2, has been permanently destroyed by the defeater.

The problematic cases, however, are those in which there is a true proposition which, when conjoined with the defeater and the confirming evidence, is such that the resulting conjunction provides a justification of p for S, but does *not* do so because the true proposition combines with all or some portion of the original, defeated confirming evidence. Such propositions merely "cancel" the effect of the defeater and restore the original justification. For example, in the Slip-of-the-Pen CPA Case, the proposition that the CPA inadvertently left out the "not" cancels the effect of the defeating proposition concerning the existence of the note. So far, so good.

Now let us return to the two Grabit Cases as presented by Barker:

	Second Version		First Version
e	There is a Tom-like-looking person removing a book from the library by concealing it under his coat.	e	The same as in the second version.
p	Tom stole the book.	p	The same as in the second version.
d	Tom has an identical twin, John, who was in the library on the day in question.	d^1	Mrs. Grabit says that d.
r	John was not in the place where the Tom-like-looking person was seen stealing the book.	r^1	Mrs. Grabit is demented and not d.

Clearly, we are to take r^1 to be a restorer of the original justification and r not to be able to restore the original justification because it has been destroyed by d.

Now, the question is: Why is r^1 a restorer and r not a restorer? Barker does not provide an answer, but we may say, *in a preliminary fashion*, that once S learns that d, S must take account of it each time he/she wishes to justify some claim about Tom on the basis of the type of perceptual evidence in e. After all, each time S sees a Tom-like-looking person, it could be John. That r is true does nothing to restore the original justification, because once we learn that Tom has an identical twin in the near vicinity we must have additional evidence beyond that in e in order to justify a belief that the Tom-like-looking person is, in fact, Tom. (More about the boundaries of the "near vicinity" shortly.)

But the situation is not the same in the first version. For in that case, the combination of d^1 and r^1 shows that we are not required to have evidence beyond the perceptual type found in e. That is, after discovering that there is no twin in the vicinity who could be mistaken for Tom (because there is no twin at all), e-type evidence, i.e., propositions representing the whereabouts of the person who looks like Tom, is perfectly adequate to justify p-type propositions, i.e., propositions representing Tom's whereabouts.

I said that this was a "preliminary" way of putting the matter, and some revision will be called for shortly, but we are now in a position to point out why r^1 is a restorer and r is not. To return to the first version—the so-called "misleading evidence case": as we have seen, the misleading defeater, d^1, defeats *only because* it renders plausible a false effective defeater (d), without employing a false member of E_s. In other words, d^1 is a misleading initiating defeater. The initiating defeater is effective only because it renders plausible another proposition—the effective defeater, d—and that effective defeater is a false proposition which occurs in the D-chain prior to any false proposition from E_s. Once it is discovered that d is false, d^1 no longer can defeat any justification based upon e-type evidence.

Thus, the original justification is restored, because the restorer, r^1, is the denial of the effective defeater upon which d^1 depends in order to be an initiating defeater. Consequently, we can say that r^1 is a restorer of the *original* justification if it entails the denial of the effective defeater upon which d^1 depends in order to defeat.

There is, of course, a slightly less complicated way of putting this point: Since d^1 is linked to an effective defeater only because it renders plausible some false proposition, we can avoid the troublesome misleading evidence cases by excluding misleading initiating defeaters from the class of propositions able to make a justification defective. That is, S's true, justified belief is knowledge if and only if all initiating defeaters of S's justification are misleading initiating defeaters. But that is exactly the analysis of knowledge which I proposed in the previous section! Thus, the distinction between misleading and genuine initiating defeaters seems necessary to explicate the notion of restorable but defeated justification.

Before considering the relative merits of Barker's proposal and my proposal for handling what he calls the "vague" cases of knowledge, I would like to consider a possible objection to my proposal, because doing so will further clarify the distinction between misleading and genuine initiating defeaters.[41] It may be thought that d, itself, even in the second version of the Grabit Case, is a misleading initiating defeater (rather than a genuine one as I claimed). For it may seem to defeat only because it renders plausible a false proposition without employing a false member of E_S, namely:

d* John was in the place in the library where a Tom-like-looking person was seen stealing the book.

In order to show that d is not a misleading initiating defeater, let us consider the three conditions which d would have to fulfill in order to be a misleading initiating defeater. They are:

MD1 d is true and not a member of E_S.

MD2 d is the first link of a D-chain which terminates in an effective defeater and which contains a false link prior to every link containing a false member of E_S. (In this case, the false link in the D-chain is d* and it is also the effective defeater.)

MD3 Every D-chain beginning with d contains a false link prior to every link containing a false member of E_S.

Roughly, MD2 asserts that d defeats *because* it renders plausible d*, and MD3 asserts that d defeats *only because* it renders d* plausible. In other words, MD3 requires that *every* effective defeater is obtained from d through a false proposition not already in E_S.

The truth of MD1 is guaranteed by stipulation in the second version of the Grabit Case. MD2 is satisfied, since e is a true member of E_S and (d&e)Rd*. But MD3 is not satisfied. That condition required that d depend essentially upon a false proposition (d*, for example) in order to be linked by a D-chain to an effective defeater. Simply put, the reason that MD3 is not fulfilled is that d is sufficient by itself to defeat the original justification, because it undermines the perceptual information contained in e. The proposition, d, is its own effective defeater, because (d&e)₵p. It is not necessary that d* be plausible, since even if it were not more reason to believe that the Tom-like-looking person is a facsimile of Tom than to deny it, the mere presence in the general vicinity of a facsimile is enough to undermine e. Thus, although there is *a* D-chain which includes a false proposition (i.e., d*) in a link prior to any link in which a false proposition from E_S occurs, d is not essentially dependent upon that or any other false proposition to lead to an effective defeater, because it is an effective defeater itself. In other words, not every D-chain contains a false proposition.

I trust that my proposal is now somewhat more clear. But this last point concerning the defeating effect of a proposition representing the presence of the twin in the "general vicinity" of Tom leads to the final comment I wish to make about Barker's alternative proposal. It concerns the necessity of portraying the inherent vagueness of the concept of knowledge by interpreting the distinction between defeated (but restorable) and destroyed justification in light of the more fundamental distinction between misleading and genuine initiating defeaters. Let us return to the second version of the Grabit Case once again. Surely there will be conflicting intuitions here concerning the requisite size of the "general vicinity" in which the Tom-facsimile is located in order for e-type evidence to be rendered defective. For example, are all, some, or none of the following genuine defeaters of justifications based upon e-type evidence?

John is *on the part of the campus* in which the library is located.

John is *on the campus* in which the library is located.

John is *in the city* in which the library is located.

John is *in the county* in which the library is located.

John is *in the country* in which the library is located.

John is *on the continent* in which the library is located.

John is *in the universe* in which the library is located.

Some of the conflicting intuitions concerning the scope of knowledge can be explained by recognizing that the source of the conflict springs from varying views concerning the size of the requisite "general vicinity" in which the facsimile is located. What I termed the "preliminary" way of interpreting Barker's proposal will suffice here. But there are other vague cases—in fact, cases cited by Barker—in which the preliminary interpretation is clearly inadequate. Without the distinction between misleading and genuine initiating defeaters, Barker's proposal provides no guidance in these cases.

Consider a variation of the first version, as Barker suggests, and "instead of supposing that Tom's mother is demented let us suppose that she is quite sane, and fabricated the twin story because she wanted to protect Tom, whom she feared might have been the culprit."[42] Does S know that Tom stole the book? I agree with Barker that there may be conflicting intuitions here—and perhaps even in the first and second versions of the Grabit Case as well—but unless we give an interpretation of restored justification in terms of misleading initiating defeaters along the lines which I have suggested, Barker's analysis does not appear to capture the source of the vagueness.

For consider the "preliminary way" of construing the distinction between defeated (but restorable) and destroyed justification. That interpretation of Barker's proposal will not capture the vagueness, since this variation of the first version shares with it the fact that, once we learn that Tom does not have a twin, the e-type evidence is again sufficient in all cases. Hence, given my *preliminary interpretation* of Barker's views, his analysis fails to capture the vagueness in these cases. For it would turn them all into cases in which the justification is restorable.

To see the merits of my proposal more clearly, consider another case in which I believe intuitions may vary. Gilbert Harman presents it in the following:

> Suppose that Tom enters a room in which many people are talking excitedly although he cannot understand what they are saying. He sees a copy of the morning paper on a table. The headline and main story reveal that a famous civil rights leader has been assassinated. On reading the story he comes to believe it; it is true; and the condition that the lemmas [premisses in the justification] be true has been satisfied since a reporter who witnessed the assassination wrote the story that appears under his byline. According to an empiricist analysis, Tom ought to know the assassination had occurred. It ought to be irrelevant what information other people have, since Tom has no reason to think that they have information that

would contradict the story in the paper. But this is a mistake. For, suppose that the assassination has been denied, even by eyewitnesses, the point of the denial being to avoid a racial explosion. The assassinated leader is reported in good health; the bullets are said, falsely, to have missed him and hit someone else. The denials occurred too late to prevent the original and true story from appearing in the paper that Tom has seen; but everyone else in the room has heard about the denials. None of them know what to believe. They all have information that Tom lacks. Would we judge Tom to be the only one who knows that the assassination has actually happened? Could we say that he knows this because he does not yet have the information everyone else has? I do not think so. I believe we would ordinarily judge that Tom does not know.[43]

Thus, Harman seems to believe that when a hitherto reliable source of information asserts something inconsistent with what was justifiably believed, then, *that, by itself,* is sufficient to defeat the original justification.[44]

I am not certain about that, primarily because of the similarity between this case and the misleading-evidence Grabit Case presented earlier. For suppose that Mrs. Grabit had convinced others that she had twin sons and that, for some reason, various newspaper and radio reports tell of her imaginary progeny when reporting the theft of the book. To paraphrase Harman: Would we say that S knew that Tom stole the book when S "does not yet have the information everyone else has"? Does the fact that many other people believe what Mrs. Grabit said turn this into a case in which S fails to know? Would one false newspaper account, written by a hitherto undetected pathological liar and believed by all those who read it, be sufficient to show that S does not know? Does the *number* of false newspaper stories make a difference? If so, suppose that Mr. Grabit, Tom's father, shares in the mother's delusions. Suppose that cousins also share in the delusion. Does that make a sufficient difference in the status of S's knowledge? I ask these questions not to disagree with Harman's evaluation of the Civil Rights Worker Case, but rather to point out that my proposal is able to capture the conflicting intuitions here, whereas Barker's proposal is not. For the preliminary way of interpreting Barker's proposal would seem invariably to lead to the conclusion that S does have knowledge in all the *vague* cases, since, once it is discovered that the second set of reports was incorrect, the *original* evidence for the proposition is again sufficient to establish it.

But my proposal is sensitive to the conflicting intuitions; for MD3 can account for them. First, it is important to note that MD3 says that an initiating defeater is misleading only if *every* D-chain contains a false link prior to every link containing a false proposition from E_S. Thus, even though an initiating defeater may anchor a D-chain in which a false proposition is rendered plausible, if it is sufficient *by itself* to defeat S's justification there will be a D-chain which does not contain a false link. The chain will begin and end with the *true* initiating defeater which is also the effective defeater. Hence, it is not a misleading defeater. As I have said, those who share Harman's intuitions would believe that 'the newspaper and radio accounts deny that the civil rights leader was assassinated' defeats the justification, by itself, and not only because it anchors a D-chain with the requisite false proposition. Similarly, if many people shared Mrs. Grabit's delusion, that fact alone *may* be sufficient to defeat the justification. That is, when a sufficient number of hitherto reliable sources of information sincerely believe a given proposition which is in conflict with what has been justifiably believed, *that fact by itself* is enough to defeat the justification. The third necessary condition of a misleading initiating defeater would not be satisfied, because the initiating defeater does not depend upon a false proposition in order to produce an effective defeater, although it may render plausible a false proposition that does defeat.

The question is: Do the initiating defeaters depend upon false propositions in the way that an initiating defeater must if it is to be misleading? Specifically, does the true proposition defeat *only* because it anchors a D-chain in which some false proposition is rendered plausible without the aid of a false member of E_S? I suggest that our varying intuitions about the answer to that question can account for our varying intuitions concerning the state of S's knowledge in the controversial or vague cases. But that consideration tends to substantiate my proposal for distinguishing misleading and genuine initiating defeaters, since this analysis can account not only for those cases in which intuitions are widely, if not universally, shared but also for those cases in which intuitions vary.

Thus, the distinction between misleading and genuine initiating defeaters must be used in order for Barker's proposal to capture successfully our intuitions concerning the vague cases. Consequently, Barker's proposal is *at best* superfluous, since it depends upon another distinction which is sufficient to distinguish between defective and nondefective justification.

But this more fundamental distinction has itself been criticized. Once again, it will be useful to consider the criticism in order to clarify the proposal further.

My proposal for distinguishing misleading and genuine defeaters has been criticized by Steven Levy in his article, "Misleading Defeaters."[45] Actually, the proposal he is criticizing contained a slightly different version of the characterization of misleading and genuine initiating defeaters, but his objection, if valid, would apply equally well to the version developed in the previous section. He claims that there are clear counterexamples to the proposal—i.e., cases in which S knows that p, but in which there is an initiating defeater which is not a misleading one. The purported counterexample which he gives is yet another variation of the Grabit Case:

> S sees Tom Grabit remove a book from the library by concealing it beneath his coat. S is ultimately going to conclude that Tom stole the book, but first he must rule out other possibilities. To this end we may suppose that S is justified in believing that Tom is not an employee of the library (S inspected the University records the day before and found that Tom was employed by the maintenance division and not the library). Furthermore, S is justified in believing that no employees of the library may remove books (such was printed in the latest bulletin). Given these beliefs, the inference to h, that Tom's removal of the book was unauthorized, is certainly warranted. Since Tom did in fact steal the book, there should be little hesitation in saying that S knows that he did. (For our purposes here we shall also suppose that Tom does not have a twin brother nor did anyone ever assert that he had.)
>
> Let us now complicate the case by adding that, on the morning in question, for reasons that only deans and vice-chancellors understand, the names of some maintenance employees, including Tom's, were transferred to the library's budget although there was no change in duties. Technically— Tom became an employee of the library. Also, the rule that no employees may remove books was altered slightly in order to allow the head librarian to examine at home a rare first edition of Rader's anthology. This change was unknown to all but a few. Although the facts stand to the contrary— for all intents and purposes Tom is not an employee of the library, and, even if he were, he would not be one permitted to remove books. S's knowledge that Tom stole the book is not impaired by the bureaucratic shuffling.[46]

Levy claims that the compound expression, let us call it d^+, "Tom is an employee of the library, and some employees of the library may remove books" is a defeater, but not a misleading one, according to my analysis.

However, the purported counterexample depends upon d^+ obscuring the fact that there are two distinct types of library employees—the t-type ("technical" type) whose names *merely appear* on the budget list and the *aip*-type (the "all intents and purposes" type) whose duties and privileges towards the library are determined by the college bulletin and administrative hierarchy. Once the *propositions* expressed by d^+ are delineated, it is clear that each of them either is not an initiating defeater at all or is, in fact, a misleading initiating defeater, according to my analysis.

The compound expression, d^+, is composed of the expressions:

a^+ Tom is an employee

b^+ Some employees of the library may remove books.

Since the meaning of "employee" is ambiguous, a^+ and b^+ express any of the following propositions:

a^+ $\begin{cases} a_1 & \text{Tom is a t-type employee.} \\ a_2 & \text{Tom is an aip-type employee.} \\ a_3 & \text{Tom is either an aip- or a t-type employee.} \end{cases}$

b^+ $\begin{cases} b_1 & \text{Some t-type employees may remove books.} \\ b_2 & \text{Some aip-type employees may remove books.} \\ b_3 & \text{Some aip- or t-type employees may remove books.} \end{cases}$

Thus, d^+ expresses any one of nine propositions:

d_1: $a_1 \& b_1$	d_2: $a_1 \& b_2$	d_3: $a_1 \& b_3$
d_4: $a_2 \& b_1$	d_5: $a_2 \& b_2$	d_6: $a_2 \& b_3$
d_7: $a_3 \& b_1$	d_8: $a_3 \& b_2$	d_9: $a_3 \& b_3$

None of these are genuine initiating defeaters. For d_1 and d_4 through d_7 are false. The proposition, d_2, does not satisfy MD2 because it does not anchor a D-chain with a false link, since it is incapable of rendering anything plausible which would lead to an effective defeater. Finally, d_3, d_8, and d_9 would anchor D-chains only if they rendered plausible the false d_1, d_5, and $(d_1 \text{ vd}_5)$, respectively. Thus, although this Grabit example is somewhat more complicated than the original Grabit Case, my proposed analysis of knowledge, with its characterization of misleading initiating defeaters, handles it correctly.

The objection just considered attempted to demonstrate that there were cases in which S *does* have knowledge, but in which my analysis portrayed the justification as defective. Now I would like to consider one from the other side, namely, the claim that there are cases in which S *does not* have knowledge, but in which all the conditions of my analysis of knowledge are fulfilled. In particular, it is claimed that the difficulty lies with the condition of defectiveness, because it counts a clearly defective justification as one which is nondefective.

The objection was developed by Kenneth Lucey and can be put as follows:[47]

> Suppose that S is well acquainted with the character traits of her business associate, Mr. Deception. She knows him to be a rogue who is constantly trying to deceive people, including S, herself. One day she meets Mr. Deception, who in the course of the conversation, asserts that $\sim h$. As it happens, Mr. Deception actually believes that $\sim h$ and says $\ulcorner \sim h \urcorner$ because he has come to realize that S is going to believe the negation of whatever he says. True to form, S does not come to firmly believe that h — which as accident would have it, just happens in this case to be true! The evidence which S has is a conjunction of e_1, *Mr. Deception has always lied to S in the past,* and e_2, *Mr. Deception is now asserting that $\sim h$.* S does not know, of course, that Mr. Deception has realized that S knows that he lies. What Mr. Deception doesn't know is that he is unintentionally leading S to the truth.

We can summarize the important facts in the case in this manner:

(1) h is true, but Mr. Deception believes that $\sim h$.

(2) $(e_1 \& e_2)Ch$.

(3) S believes firmly that h.

Thus, we can assume that S has a true, justified, and firmly held (psychologically certain) belief that h. But, as Lucey correctly points out, S does not possess knowledge that h. Since three of the four conditions of knowledge are fulfilled, namely, K1 through K3, if my characterization of defectiveness is correct, there must be a genuine initiating defeater.

But what could it be? Let us consider first whether it is something like 'Mr. Deception really believes that $\sim h$ and said $\ulcorner \sim h \urcorner$ because he knows that S has discovered him to be the liar that he is'. Call that proposition 'd_1'. Is d_1 a genuine initiating defeater?

For reasons similar to those discussed in connection with the misleading-evidence Grabit Case and the Civil Rights Worker Case, I suspect that intuitions will vary here. Some may believe that d_1, alone (like the propositions representing the false reports in the Civil Rights Worker Case and Mrs. Grabit's concocted story designed to save her son), is an *effective* defeater as well as an initiating defeater. Thus, they would hold that there is a D-chain beginning with d_1 and ending with an effective defeater which does not contain a false proposition. The chain is simply '$d_1 R d_1$'. As mentioned before, since R is reflexive, such chains are permissable.

But the parallel with the Grabit and Civil Rights Worker Cases suggests that others may believe that the initiating defeater, d_1, is not effective unless it first renders plausible the false proposition that $\sim h$. In other words, they would hold that the D-chain must contain a false proposition prior to any link which contains a false member of E_S. In short, d_1 could be seen as a misleading initiating defeater.

There are two points to consider regarding the varying intuitions concerning whether d_1 is a genuine or a misleading initiating defeater. The first is that those who believe that d_1 is an initiating as well as an effective defeater will also grant, presumably, that d_1 renders $\sim h$ plausible. Now, since $\sim h$ is false, it might be thought that d_1 would thereby become a misleading rather than a genuine initiating defeater. But if it is correct that d_1 is an effective defeater, the situation here is similar to the Civil Rights Worker Case in which the initiating defeater did render plausible a false effective defeater, but there is nevertheless a D-chain terminating in an effective defeater which does not contain a false link, namely, the one mentioned above, $d_1 R d_1$. That d_1 also renders plausible a false effective defeater, $\sim h$, shows only that there is one D-chain which satisfies MD2. It does not show that *every* D-chain is such a chain. And if d_1 is a misleading initiating defeater, MD3 requires that *every* D-chain contain some false proposition. Hence, the intuition that the justification is defective because of d_1 is consistent with my characterization of misleading and genuine initiating defeaters.

The second point to note concerning the varying intuitions is this: Since even those who think that d_1 is not a genuine initiating defeater—because it must first render plausible the false proposition,

~h, in order to become effective—will presumably believe with Lucey that S's justification is defective. Hence, if my proposal is to satisfy them, it must demonstrate that there is a genuine initiating defeater. Now, as Lucey pointed out, it cannot be the proposition *Mr. Deception is wrong about ~h*, since, even though that proposition satisfies MD1—because it is true and not a member of E_S—if it is conjoined with $(e_1 \& e_2)$ it continues to confirm h! An effective *defeater*, of course, must be such that the conjunction of it with the evidence *fails* to confirm h.

Nevertheless, there is a clear, genuine initiating defeater. To see what it is consider a more simple case in which S relies on the testimony of a hitherto reliable person, Ms. Verity. Suppose that Ms. Verity asserts that p. Just as we did in the Mr. Deception Case, let us grant that the conjunction of e^1, *Verity says ⌜p⌝*, with e^2, *Verity says what she believes to be the truth*, confirms p; that is, $(e^1 \& e^2)Cp$. Now, if Ms. Verity happens not to know that p (even though p is true), S's justification is defective. The reason is that there is a genuine initiating defeater, namely, *Ms. Verity does not know that p*. Call that proposition, 'd^1'. That is, $[d^1 \& (e^1 \& e^2)] \not\!C p$. For d^1 renders Ms. Verity's testimony completely useless. Note that d^1 does not render ~p plausible, since it is mute with regard to the truth of p. In fact, whether p is true or whether it is false is irrelevant to the defectiveness of S's justification for p by $(e^1 \& e^2)$. What makes it defective is simply that Ms. Verity fails to know whether p is true.

The same principle applies to the case suggested by Lucey. The truth or falsehood of h or the fact that Mr. Deception's belief is false is not what makes the justification *clearly* defective. What is crucial is that Mr. Deception did not know whether h was true or whether it was false. S cannot rely upon his testimony, even if S attempts to compensate for his previous prevarications. The proposition *Mr. Deception does not know whether h is true or whether h is false* is true and not a member of E_S. It is also such that, when conjoined with $(e^1 \& e^2)$, the resulting conjunction no longer confirms h. In addition, that proposition does not render plausible any false propositions, much less a false link of a D-chain prior to every link containing a false proposition from E_S. Hence, it is both a genuine initiating defeater and an effective defeater. In short, there is a proposition which makes the justification defective according to my suggested characterization of defective justification.

Consequently, I believe that my proposal correctly characterizes the Mr. Deception Case. As some may believe, S would lack knowledge because Mr. Deception believes that ∼h and has discovered that S knows about his habit of deceiving others. They can appeal to my characterization of defectiveness for support of their intuitions. But, be that as it may, S *clearly* lacks a justification immune from attack because Mr. Deception does not know whether p is true. S's justification is clearly defective because of the genuine initiating defeater representing that fact. Thus, all the relevant intuitions, even conflicting ones, can be captured by my proposed characterization of knowledge.

Summary of this section

It will be recalled that the purpose of considering what I take to be the best alternative to my method of characterizing defective justification and of defending my proposal against the most plausible objections was to clarify my account. That was necessary before beginning to determine whether that account is adequate for providing a basis for an analysis of absolute certainty. That issue will be before us shortly. But first, there is one other extremely relevant objection which must be disposed of. It is that my proposal and defeasibility theories in general lead to what I have called Direct Scepticism—the brand of scepticism which claims that S cannot know any of those propositions which are normally believed to be knowable. We had better determine whether that objection is correct before using my analysis of knowledge as part of a refutation of Direct Scepticism!

3.10 A Particularly Relevant Objection to the Defeasibility Theory: Does It Lead to Scepticism?

My proposal for an analysis of knowledge has withstood the various criticisms thus far considered, but the crucial test of this proposal, as far as scepticism is concerned, lies in its ability to satisfy the six desiderata of an acceptable account of certainty. Before putting it to the test, however, I would like to consider yet one more, particularly relevant objection which has been raised against all defeasibility accounts. It has been claimed that, far from it being the case that the defeasibility account provides a basis for rejecting scepticism, all defeasibility accounts lead to Direct Scepticism!

I believe not only that locating the errors in this objection will help to clarify further the defeasibility account in general, but that a discussion of the issues raised by the objection will also prepare the way for a consideration of the ability of the defeasibility theory to characterize certainty in a way which does not lead to scepticism.

In "Defeasibility and Scepticism,"[48] Robert Almeder argues that any properly constructed defeasibility theory will inevitably lead to scepticism. The argument presented by Almeder, although designed with three particular defeasibility theories in mind,[49] is easily generalizable to apply to all such theories. The common feature of all defeasibility theories, according to Almeder, is that they include a claim that a justification for p is defective "if there is some true sentence such that if it were added to a person's justification [for p] he would not be justified in believing what he does believe."[50] We have seen that, because of the existence of misleading information, restrictions must be placed upon the class of true sentences which render a justification defective. As will be apparent when we examine Almeder's argument, however, it could easily be redesigned to accommodate this modification by confining his claim to the set of the appropriately restricted class of true propositions, which makes a justification of p defective — i.e., the class of genuine initiating defeaters.

Almeder claims to have demonstrated two points:

A1   There are counterexamples (other than the ones already considered) to defeasibility analyses of knowledge; and

A2   In order to avoid these counterexamples, defeasibility analyses must require that any justification of p on the basis of some evidence, e, must be such that e entails p.

Almeder is certainly correct that the adoption of the requirement 'eCp only if e → p' would lead to Direct Scepticism. But I believe that the purported counterexamples fail and, equally important, that, even if they were genuine, A2 is false.

Let us look at the purported counterexamples first. A recipe for constructing them is given by Almeder in the following rather long quotation:

> . . . suppose that there is a true sentence [in the appropriately restricted class] which would defeat a person's justification if it were added to his justification, but that only God knows that the sentence is true. Moreover, God . . . has effectively decreed that the race of men unto

eternity shall never be allowed to know that this sentence is true. . . .
Here, then, we should have a statement which is both true and such that
if it were conjoined to the justification the justification would be defeated. The condition of non-defectiveness . . . is not satisfied, and so,
assuming the condition correct, we must say in this case that the race
of men do not know what they claim to know. . . . But as a matter of
fact, would we *say* that the race of men do not know what they profess
to know if what defeats their claim could never in fact be known by men?
I think not, and the reason is simple. As long as a person claims to know
that p, and we cannot ever know what would defeat the claim, we must
and so *say* he knows. As Austin says, it is outrageous to *say* that he might
not know what he professes to know unless we have some definite lack in
mind, one which we are prepared to specify upon being pressed. We cannot go on *saying* 'But that's not enough' unless we are willing to state
specifically what would be enough. [Emphasis added.] [51]

To demonstrate that this recipe is unsatisfactory, consider the results of applying it in the following cases. Suppose a person, S, "professed" to know something, h, on the basis of some evidence, e,
where e confirms h, but where e is false. Suppose further that, owing
to divine decree, it will never be known that ~e. According to the
recipe, we must *say* that S does know that h even though e is false.
To make matters worse, suppose further that h is false, but confirmed by e. Again, S professes to know that h; but ~h is forever hidden
from human knowledge. Now, if Almeder were correct, we should
say that S knows that h, if he/she professed to know that h, even
though both h and e are false!

Obviously, something has gone wrong. It seems to me that Almeder has conflated the distinction between those occasions when we
justifiably *say* (believe, think, affirm, etc.) something (in this case,
'S knows that h') and those occasions when *what we say* is true. In
other words, if this recipe is to cook up counterexamples, it must
produce a case in which the proposed conditions of defective justification are fulfilled and yet S does know that p. Instead, the recipe
only illustrates situations in which the conditions of defective justification are fulfilled but (since we are prevented from knowing that
fact) we are *justified in saying* that S does know that h.

Almeder is correct in believing that if the genuine initiating defeater is forever hidden from human view, it would be "outrageous"
to *say* (believe, think, affirm) that S does not know what he/she

professes to know. There would be no evidence available for such a claim. But unjustified propositions, even "outrageous" ones, are sometimes true. To put it another way, nothing prevents a proposition from being false and justified; in this case, the false, justified proposition is *Ksp*.

The argument for the second claim is as follows:

> The counter example has shown that a . . . defeater can defeat the claim only if the defeater is humanly knowable. . . . The only thing we could do to save the condition of non-defectiveness would be to require that whatever defeats the justification be humanly knowable. But it will not help to incorporate this requirement into the proposed condition of non-defectiveness. For . . . there cannot be no unknowable defeater of a justification unless the justification is an *entailing* one. . . . As urged above, as long as it is logically possible that there be an unknowable defeater, a person's justification can be defective under the proposed condition even though it be possible that he knows what he claims to know. . . . It should now be clear that *that* possibility is not blocked by adding to the original condition that what defeats a justification must not be unknowable: it is not blocked unless the justification is entailing.[52]

Now, insofar as this argument depends upon the authenticity of the purported counterexample, it has already been shown to be unsound. Be that as it may, it still does not follow that, in order to avoid the counterexample, we must require that the justification is entailing.

First, it is not even true that, by making the justification entailing, we eliminate the possibility of an unknowable defeater. For, following our earlier example, suppose that e confirms h and (as per Almeder's suggestion) that e entails h. But, again, suppose that e is false. In addition, let ~e be "humanly unknowable." Surely, the true proposition, ~e, is a genuine initiating defeater; hence, there is an unknowable defeater even though e entails h.[53]

What is more important, however, is Almeder's claim that the possibility of an unknowable defeater cannot be excluded by incorporating the requirement that all defeaters be knowable into the condition of nondefectiveness, because he believes that there would remain the logical possibility of an unknowable defeater.

Almeder seems to believe (falsely) that the counterexamples can be blocked only if the proposed conditions of knowledge eliminate the *logical* possibility of unknowable propositions serving as defeaters. Now, assuming that the purported counterexamples were genuine, Almeder would be correct that we cannot block the *logical* possibility of there being such propositions by requiring that all defeaters be knowable. But it is not required that the *logical* possibility of unknowable defeaters be removed in order to block the counterexamples. For a justification is not defective because of the mere logical possibility of genuine defeaters. Rather, *a justification is defective only if there actually is a genuine defeater.* The mere *possibility* of a true, unknowable proposition would not create a situation in which S knows that h and the condition of defectiveness is fulfilled. Only an actually true, unknowable proposition would create such a situation. In other words, S's justification is defective only if there is an actual genuine defeater. The *logical possibility* of a genuine defeater only makes it logically possible that the justification is defective. Thus, there is no reason to block the logical possibility that the justification is defective.

I suspect that Almeder may have conflated the following two claims:[54]

(1)   If S knows that p, then there is no logical possibility of there being an unknowable defeater.

(2)   There is no logical possibility both that S knows that p *and* that there is an unknowable defeater.

The former, (1), is clearly false, and the latter, (2), is true. That (1) is false can easily be seen by comparing it with:

(3)   If S weighs 200 pounds, then there is no logical possibility that S weighs less than 200 pounds.

The contingent claims in the antecedents of (1) and (3) do not imply the modal statements in the consequents. (Imagine the dismay of someone who weighed 200 pounds and was attempting to lose weight, if (3) were correct!) But if Almeder thought that (1) were true, his belief that the defeasibility analyses must eliminate the *logical possibility* of unknowable defeaters could be explained. For if (1) were true, and S did know that p, it would follow that it is not logically possible that there is an unknowable defeater. But, to repeat, the *logical possibility* of defeaters need not be eliminated; only the logical possibility of it being the case that S knows that p and there is a defeater (knowable or not) of S's justification for p.

Thus, even if the recipe for counterexamples did, in fact, produce counterexamples, just as the defeasibility analysis can incorporate the condition that misleading initiating defeaters do not make a justification defective, the analysis could be adjusted to incorporate the condition that unknowable defeaters do not make a justification defective.[55] That could be done, of course, without requiring that e entail p.

I trust that it is clear that Almeder has not shown either that there are counterexamples to the defeasibility account or that, in order to avoid those counterexamples, the defeasibility analysis must embrace Direct Scepticism. The path should now be clear to test my proposed model of knowledge against the six desiderata in order to determine whether the conjunction of K1 through K4 meets the requirements of the sceptic with regard to absolute certainty.

3.11 D1 and D5 Are Satisfied: Knowledge Clearly Entails Evidential and Psychological Certainty

In Chapter Two I argued that there are no good reasons for accepting the claim made by the Direct Sceptics that no person can have knowledge. In the present chapter I have developed a model of knowledge in which to test my claim that there are good reasons for believing that many persons know many things. That can be demonstrated if there are good reasons for believing that all the necessary conditions of knowledge are jointly fulfilled. I have granted that one of those necessary conditions of knowledge is absolute certainty. For without granting that to the sceptics, they could rightfully claim that my proposed refutation avoided the core of their position.

In order to show that the concept of certainty presented here —namely, nondefective justification—is acceptable in general and, in particular, that it satisfies the demands of the sceptics, we delineated six desiderata of an adequate account of evidential certainty. I believe that we are now in a position to determine whether the model of knowledge developed in sections 3.8 through 3.10 meets the requirements specified by the six desiderata. That it fulfills some of them is relatively easy to show, requiring not much more than a few sentences. To show that others are fulfilled will require more than one section.

Let us begin with two of the desiderata which are rather obviously fulfilled by my account. The fifth one requires that the entailment between knowledge and evidential certainty be manifest. Since my claim is that evidential certainty is nothing other than the conjunction of K3 and K4, the proposition *S knows that p* entails the proposition *p is certain for S*. In addition, the first desideratum is satisfied if the proposal provides us with a way of maintaining the distinction between evidential and psychological certainty. That this is possible is apparent since the belief condition, K2, in the proposed analysis of knowledge, and the accompanying requisite degree of tenacity are independent of the two conditions, K3 and K4, which are jointly sufficient to render a proposition evidentially certain. As we will see shortly, my proposed analysis of absolute evidential certainty can specify obtainable limits toward which propositions can become increasingly (relatively) evidentially certain. Thus, Ayer's view that S knows that p only if S has the (evidential epistemic) right to be certain (psychologically) that p and the implicit relationship between evidential and psychological certainty can be clarified. For S would have the right, whether absolute or relative, to such tenaciously held beliefs, depending upon how close S's evidence for p approaches the requisite conditions of absolute evidential certainty. We have also seen that, contrary to the views of some philosophers, neither absolute psychological certainty nor absolute evidential certainty lead to dogmatism (see sections 2.11 and 3.5).

3.12   D2: Nondefective Justification and Absolute Certainty

The previous section showed that we can count D1 and D5 as satisfied. So much for the desiderata which are easily fulfilled! Unfortunately, the arguments to show that the other desiderata are satisfied are more complex and controversial. I will postpone considering D6 until we have completed the discussion of D2, D3, and D4.

Now, since the primary purpose of Chapter Three is to show that many contingent, empirical propositions are absolutely certain, the success of my characterization of certainty depends essentially upon its ability to fulfill D2—the requirement that the concept of absolute certainty be explicated in a way which satisfies the reasonable demands of the sceptic. If that can be done, an adequate characterization of relative certainty and its relationships to absolute certainty can be provided comparatively easily. For both can be understood by employing the analysis of absolute certainty. Consequently, the discussion of the ability of my proposal to meet D2 must proceed carefully.

That discussion will be presented in three stages. In this section, I will review briefly the two features of absolute certainty claimed by the sceptic and show that nondefective justifications, if there be any, would satisfy those requirements. In the next section, I want to demonstrate that many empirical, contingent propositions *can* satisfy those requirements. Finally, in a subsequent section, 3.16, I will argue that many empirical, contingent propositions do, in fact, satisfy those requirements.

Let us begin by reiterating the two requirements of absolute evidential certainty demanded by the sceptic:

EC1   If p is absolutely certain for S, then p is justified for S, and there are no grounds which make p at all doubtful for S.

EC2   If some proposition, say i, is more certain than p, then p is not certain.

My claim is that EC1 and EC2 are satisfied by a proposition if that proposition is nondefectively justified.

Let us begin with EC1. If a proposition is nondefectively justified, then there are absolutely no genuine initiating defeaters of that justification. The justification is, so to speak, attack-proof.

Now, it is important to be clear about what I am claiming here— *and* what I am not claiming. I *am* claiming that, if there are no genuine initiating defeaters of S's justification of p, then there are no legitimate reasons which would justify S in giving up the belief that p. But I am *not* claiming that:

(1)   There are no defeaters of the justification; or

(2)   S could not be induced by hook or by crook to give up the belief that p; or

(3)   S could not come to doubt p.

None of those claims is a requirement for evidential certainty as specified by my proposal. First, there may be *misleading* initiating defeaters of the justification; but they are not legitimate grounds for doubting p. For, if the third condition of a misleading defeater is fulfilled, that is, if MD3 is fulfilled, the defeating effect of the initiating defeater will depend essentially upon a false proposition in the D-chain. Second, various events could occur which, no doubt, would cause S to give up the belief that p. S could be drugged, could be misled by deliberately planted misleading information, or could come to accept a false proposition, f, which is such that $(f\&e)\mathbb{C}p$. Finally, the claim here is not that nothing could induce doubts, only that, if these doubts were to occur, they would be either the result of something unrelated to the evidence for p or the result of accepting false or misleading information.

Further, I am not claiming that, if p is evidentially certain, there *could* be no legitimate grounds for doubting that p. As we have already seen in section 3.2, such a requirement leads to the consequence that the evidence for p entail p. According to my proposal, a justification would be sufficient to provide certainty if it *is* absolutely reliable in the actual situation in which it is employed; it cannot be further required that it be able to withstand all possible defeaters.[56] What is required is that there be (actually) no significant counterevidence whatsoever.

There is one objection to my claim that EC1 is satisfied by nondefectively justified propositions which should be considered here. For it may be recalled that I chose to develop the concept of nondefective justification by proposing a "middle course" between the "weak" and "strong" characterizations of certainty presented by Firth (see section 3.2). Thus, it may appear that there is a set of more stringent requirements for certainty which is available, and, if we are to continue to grant to the sceptic everything which can be granted, we should have chosen the strongest plausible characterization of certainty.

The three characterizations of certainty mentioned in section 3.2 are:

Weak Characterization	p is certain on the basis of evidence, e, if and only if there is no true, nonmisleading proposition, d, such that $(d\&e)C{\sim}p$.
Middle Course	p is certain on the basis of evidence, e, if and only if there is no true, nonmisleading proposition, d, such that $(d\&e)\mathbb{C}p$.
Strong Characterization	p is certain on the basis of evidence, e, if and only if there is no true, nonmisleading proposition, d, such that $(d\&e)$ weakens the confirmation of p.

To restate the objection: the sceptic should not agree that EC1 is fulfilled by my analysis of certainty and should insist that a proposition is certain only if there is no new information which would legitimately weaken the justification.

When we first discussed the three characterizations of certainty, I suggested that, appearances to the contrary, the so-called middle course was in fact no less stringent than the strongest characterization of certainty, because of the nature of the concept of confirmation used throughout Chapter Two and this chapter. Now is the appropriate time to explain that remark.

Suppose some evidence, e, confirms a proposition, p, and that there is some further information, c, which is counterevidence for p. We can say that *c is counterevidence for p* if and only if c assigns a probability of greater than zero to $\sim p$. Thus, counterevidence for p may be strong enough to confirm $\sim p$, it may only render $\sim p$ plausible, or it may be so weak as merely to make $\sim p$ a real possibility. As will emerge shortly, it is necessary to include such extremely weak evidence against p in the counterevidence for p in order to accommodate the most reasonable, stringent interpretation of the sceptic's intuition, mentioned above, namely, that if S knows that p on the basis of e, then e must guarantee the truth of p.

When c is conjoined with e, there are two possible results. The proposition, c, may remain evidence against p even in the context of e. On the other hand, c may combine with e in such a way as no longer to count against p.[57] Let us call evidence against p which is rendered ineffective when combined with e 'counterevidence for p which has been *absorbed* by e.' We will call evidence which, even in the context of e, continues to count against p '*unabsorbed* counterevidence.' Counterevidence can be more or less absorbed by e, but when I use the expression 'counterevidence absorbed by e' I mean to refer only to counterevidence which has been *completely* absorbed by e so that it no longer counts *at all* against p. Any counterevidence which is not completely absorbed I will treat as unabsorbed counterevidence.

An example may help to clarify the distinction between absorbed and unabsorbed counterevidence. Consider two sets of evidence for the proposition, p, *that Deadly is a murderer.*

Evidence Set A (e_A)	Evidence Set B (e_B)
Ten reliable witnesses claim to	The evidence, e_B, contains e_A with the

have seen Ms. Deadly kill Mr. Victim. Each of them had a clear view of the killer, knows Deadly fairly well, and describes the crime in a way which is congruent with the other accounts.

addition that each of the witnesses claims to have seen another person, Mr. Accomplice, hand the murder weapon to Deadly and help to dispose of the body.

Let us assume that e_A Cp and e_B Cp. Now, consider a proposition, c, which is: *Mr. Accomplice says that he was standing very close to the murderer and that, although the murderer looked like Deadly, it was not Deadly.* I think that it should be clear that, although c is counterevidence for p, if e_B Cp, then $(e_B \& c)$Cp because the counterevidence, c, has been completely absorbed by e_B. In fact, given e_B, one would expect that Mr. Accomplice would point the finger of suspicion away from Deadly unless he were suddenly overcome by a guilty conscience. On the other hand, even in the context of e_A, the counterevidence continues to count against p; c is not completely absorbed by e_A.

There are two possible ways in which one might want to interpret the evidential force of c when conjoined with e_A, depending upon the characterization of confirmation employed in the analysis of justification. Some may hold that p is still confirmed although the confirmation has been "weakened" by e. From that point of view, a proposition would be confirmed whenever a clear *preponderance of the evidence* was in its favor. However, others may require more of confirmation and hold that, because c was not absorbed (completely) by e_A, the conjunction $(e_A \& c)$ does not confirm p. To extend the legal analogy, a proposition would be confirmed by evidence only if the evidence made the proposition *beyond any reasonable doubt.* From this perspective, unabsorbed counterevidence, however slight, furnishes such a reasonable doubt. It is important to note that c is *not* a misleading initiating defeater if $(e_A \& c)$Cp, because the D-chain from the initiating defeater, c, to the effective defeater, c, does not contain a false proposition. That is, in this case, the initiating and effective defeater are identical. (It may be true that in some cases c renders a false proposition plausible which is also an effective defeater, but, in order to be a misleading defeater, as we have seen in section 3.8, every D-chain beginning with c and ending with an effective defeater must contain a false proposition.)

Those who take confirmation to be a matter of degree will, presumably, argue that a proposition is absolutely certain only when there is no true, nonmisleading counterevidence which weakens the confirmation. Thus, they would select the "strong" characterization of certainty. On the other hand, those who think that a proposition is not confirmed by evidence which contains some unabsorbed counterevidence will select the so-called "middle" course. The strong characterization would be inappropriate since it implies that confirmations can be "weakened" rather than destroyed by unabsorbed counterevidence. So-called "weak" confirmations are not adequate to produce knowledge.

The crucial point to note here is that there is *no real disagreement* concerning the extension and nature of the boundary between evidentially certain and noncertain propositions. For those who choose the middle course and those who choose the strong version agree that it is the existence of unabsorbed counterevidence which prevents a proposition from being absolutely certain.

I chose to take the strict view with regard to confirmation for two reasons. First, it is the clearest way of satisfying the sceptic's requirement that it be as difficult as possible to confirm a proposition and as easy as possible to override or defeat it. This should allay the fear underlying the objection which we are considering that I may have forsaken the general principle of granting to the sceptic whatever reasonably can be granted. Second, I believe, with the sceptic, that there is an important distinction between evidence which confirms a proposition for the sake of knowledge and evidence which renders a proposition the most reasonable hypothesis to adopt. That distinction is more easily lost if the relative concept of confirmation is adopted. The reasons for this will emerge in section 3.15 when we consider the Lottery Paradox.

Thus, I will continue to describe a proposition, p, as nondefectively justified if it is justified and there is no genuine initiating defeater. This proposal is no less stringent than one which employs the "strong" characterization of certainty and the relative characterization of confirmation.

I will take it, then, that EC1 can be fulfilled by my proposal. If a proposition, p, is nondefectively justified, it is such that there are no grounds which make p doubtful. Now, what about EC2? EC2 requires that if i is more certain than p, p is not certain. That desideratum is satisfied by my proposal, for, if p is less certain than i, then there would be at least one genuine defeater of the justification of p. Consequently, p would not be certain. That is, if p is evidentially less certain that i, either i is uncertain, and then there would be at least one defeater of p (though not necessarily the same one that defeats the justification for i); or i is certain and has no genuine initiating defeaters. In the latter case, the only way a justification of p could be *less* certain that the one for i would be that the justification for p has at least one genuine initiating defeater.

To see this, compare "certainty" with the absolute term "flat," as suggested by Peter Unger.[58] If some surface, S_1, is less flat than another surface, S_2, then S_1 must have at least one bump. For, if S_2 were flat (absolutely), then S_1 would have to have at least one bump (or else it would be equally as flat as S_2); and, if S_2 were not absolutely flat (i.e., if it had at least one bump), S_1 would have to have at least one bump to be *more* flat than S_2, since it would have to have *at least* one bump to be *equally* as flat as S_2.

Thus EC2 is formally satisfied by my proposal. If i is more certain than p, p is not certain, for there is at least one genuine initiating defeater of the justification of p.

3.13 Empirical, Contingent Propositions Can Be Evidentially Certain

In the previous section, I argued that if a proposition, p, was non-defectively justified, it would satisfy the two requirements of an absolutely evidentially certain proposition—EC1, which required that if p is certain, then there are absolutely no grounds for doubting it; and EC2, which required that there be no proposition, i, such that it is more evidentially certain than p. I did not claim that any propositions are, or even could be, nondefectively justified. In fact, I suspect that the sceptic will argue that no empirical, contingent proposition could be certain. For he/she would believe that there are propositions which are less susceptible to defeat than any of those propositions which I claim are knowable. And if there are such propositions, then no empirical proposition can be absolutely certain. Hence, the sceptic will eagerly accept EC2 as a necessary requirement of absolute certainty.

Among the obvious candidates for the more certain proposition are the following:

(1)   Some contingent proposition other than p for which we have "more" evidence than we have for p. 'I exist' would be an example. It would be claimed that such a proposition is more certain than 'Jones owns a Ford'.

(2)   The evidence for p, say e, because it provides the evidential basis for p.

(3)   A tautology; say for example, p v ∼p.

For the sake of simplicity I will consider only the last candidate, since it seems the most likely to succeed. If 'p v ∼p' can be shown to be no more evidentially certain than 'p', the arguments can be easily transferred to the other candidates.

In this section, I will assume that p can satisfy K3 — the requirement that p be justified by e. The issue for this section is whether p v ∼p is *necessarily* more certain than p. In section 3.16, I will present my reasons for believing that p does satisfy K3 and K4. The task of this section is to show that p *can* satisfy K4.

Well, is 'p v ∼p' necessarily more certain than 'p'? Surprisingly, my answer is "no," *because there may be no more genuine initiating defeaters of* p *than there are of* p v ∼p.

Of course, it is *logically possible* that there are genuine initiating defeaters of the justification for p, and it is not logically possible that there are defeaters of the justification of p v ∼p (assuming that S's evidence for p v ∼p includes the evidence that this proposition is a tautology). That is, it is logically possible that p is not evidentially certain, and it is *not* logically possible that p v ∼p is not evidentially certain. But that is beside the point. The issue is whether p must be less evidentially certain than p v ∼p, not whether it is *logically possible* that it is less certain.

Nevertheless, recalling the discussion of relative certainty and, in particular, the passage quoted from Frankfurt's article, it might be argued as follows: Surely one would be willing to risk more on p v ∼p than on p. It would be a better bet. In that case, p v ∼p is more certain than p (see section 3.5).

I think that it should be granted that p v ∼p is a better bet; but it does not follow that it is more evidentially certain than p. First, this objective apparently confuses psychological and evidential certainty. For it appears that what is meant is that the degree of confidence in 'p v ∼p' is higher than the degree of confidence in 'p'.[59] That degree is relative to the willingness to take risks on the two propositions. Thus, this reply does not present any reason to believe that 'p ∨ ∼p' is *evidentially* more certain than 'p'. If it showed anything at all, it would show that 'p ∨ ∼p' is always psychologically more certain than 'p'. But it does not even show that. As argued earlier, a person may feel absolutely certain about something for causes completely un-related to the probability of its truth or the degree of evidence which he has for it.

But there is a second and more important issue at stake here. Let us grant that p ∨ ∼p is always a better bet than p. It does not follow that p ∨ ∼p is evidentially more certain. That it is a better bet can be explained by pointing out that it is always easier to *show* that one has won the bet. That is, as soon as the sentence expressing p ∨ ∼p is fully understood (i.e., as soon as it is recognized that the sentence ex-presses a tautology), a winner of the bet is obvious. A judge, whose responsibility it was to determine a winner, would have no diffi-culty in making such a determination. Thus, it is easier to *win* the bet on p ∨ ∼p than on p, but that is not because p ∨ ∼p is evidentially more certain. It is rather because, as soon as the terms of the bet are understood, the winner is obvious. Thus, it is a better bet because it is easier to *win* — and that, after all, is the purpose of betting.

Another reason which the sceptic might offer for the claim that 'p ∨ ∼p' is evidentially more certain that 'p' is that we are in a better position to determine that we know that p ∨ ∼p than we are in a posi-tion to determine that we know that p. After all, S's justification for p ∨ ∼p (again assuming that S recognizes that it is a tautology) is sufficient to justify the belief that there are no defeaters of the justification. But S's confirming evidence for p need not be sufficient to justify the proposition that there are no defeaters. For, as I have argued, the Defeater Consequence Elimination Principle is false (see sections 2.12 and 3.2). S may be justified in believing that p without being justified in believing the denial of every defeater of that justifi-cation. Thus, with regard to 'p ∨ ∼p', if K3 is fulfilled, K4 is fulfilled; whereas with regard to 'p', that relationship between K3 and K4 does not obtain. Put another way: S has adequate confirming evidence for the claim that p ∨ ∼p is evidentially certain whenever S has adequate confirming evidence for p ∨ ∼p. The same is not true of S's evidence for p.

But, once again, this is beside the point. It can be granted that S is in a better epistemic position with regard both to knowing that he/she knows that p v ~p and to knowing that p v ~p is evidentially certain than he/she is with regard both to knowing that he/she knows that p and to knowing that p is evidentially certain. But that does not put S in a better epistemic position with regard to p v ~p than with regard to p. That is, it might be easier to *determine* that we are in the best epistemic position with regard to p v ~p than it is to determine that we are in the best position with regard to p. But that does not make the position epistemically better with regard to p than it is with regard to p v ~p. It only makes it better with regard to *knowing* that p v ~p than it is with regard to *knowing* that p. Thus, I am willing to grant that it is easier to know that we know that p v ~p than it is to know that we know that p. That, however, does not imply either that p v ~p is evidentially more certain than p or that we cannot know that we know that p.

But the sceptic may wish to argue that the justification for p v ~p is more immune from attack, i.e., more secure, than the justification for p. For surely there are more possible defeaters of the justification of p by e than there are for the justification of p v ~p. Thus, p v ~p is more worthy of belief or, simply, more certain than p, if the justification of p v ~p includes the claim that p v ~p is a tautology.

Let us grant that the sceptic is correct about one thing. No matter how the world *could be*, i.e., no matter what else could be true, the justification of p v ~p is secure. On the other hand, the justification of p by e is not secure against all possible genuine defeaters.[60] In some possible situations, the justification for p will be defective; but there are no possible situations in which the justification for p v ~p is defective. Thus, the sceptic might continue, p v ~p is more certain, because there is no "chance" (to return to the way Lehrer put it— see section 3.6) of S's being wrong that p v ~p, whereas there is a chance that S is wrong that p. As long as there is such a chance, the sceptic would claim, e would not make p evidentially certain for S. To return to the claim discussed earlier, p v ~p is a better bet than p, because there is no chance that the former will lose. In other words, if e (alone) does not guarantee the truth of p, then p is not certain.

We have arrived at what I take to be the core of the sceptic's view with regard to absolute evidential certainty. That fundamental claim is that the proposition, p, is evidentially certain on the basis of e only if its truth is guaranteed by e, and the proposition, e, can guarantee the truth of p only if there are no possible defeaters of the justification of p by e.

I believe that there is a clear and convincing response to this argument for the claim that p v ~p is necessarily more certain than p. For the fact that there are more *possible* defeaters of one justification than there are of another does not make the former less certain; *it makes the latter certain in more situations.* It is true that 'p v ~p' is certain in all *possible* situations; but that does not make 'p v ~p' more certain in the *actual* situation. According to my proposal, whether our justifications provide evidential certainty depends upon whether there *is* any genuine evidence which *could* be brought against the proposition; it does not depend upon the mere possibility of there being such counterevidence. That is, if there *are* any genuine initiating defeaters, the justification does not provide for certainty. But it is not further required that there be no possible defeaters. Thus, although some justifications are certain in *more situations* than others, that fact does not make them *more certain* in any particular situation.

The distinction between *certainty in more situations than the actual one* and *greater certainty in the actual situation* is crucial to my claim that a proposition, p, can be rendered absolutely certain on the basis of e without e entailing p. Consequently, some further account of it is required.

Consider two justifications of the proposition, p, based on evidence, e_1 and e_2, respectively, and suppose that the set of possible genuine initiating defeaters of the justification of p by e_1 is a proper subset of the possible genuine initiating defeaters of the justification of p by e_2. Also, assume that there are no (actually true) genuine initiating defeaters of either justification. My claim is that, although both e_1 and e_2 render p absolutely certain in the actual situation because there are no genuine initiating defeaters of either justification, there is an important sense in which the justification of p by e_1 is "stronger" than the justification of p by e_2. It is stronger because the justification is resistant to defeaters in more possible situations.

If the sceptic were unwilling to grant the distinction which I am urging here and insisted that p is absolutely certain on the basis of e only if that justification were resistant to all *possible* defeaters, he/she would be requiring that e entail p. For if e did not entail p, then it would be possible for e to be true and p false. But, in such a situation, there would be a defeater of the justification of p by e, namely ~p. Hence, if e renders p certain only if e is logically incompatible with all possible defeaters of the justification of p by e, it must not be possible that p is false when e is true—in other words, e would entail p.

It may seem, however, that there is a course available to the sceptic which neither leads to the consequence that e entails p nor limits defeaters to the class of true propositions. For there are subsets of the possible defeaters including but not limited to the set of actual defeaters. The sceptic might require that a justification be resistant to all the possible defeaters in such a subset of the possible defeaters.

Consider the Grabit Case once again. Suppose that Tom Grabit (the real thief) does, in fact, have a twin who really might have been in the library, because he lives "in the vicinity" of the library. The sceptic may require that, in such a situation, evidence which renders p certain would have to contain information concerning the twin's actual whereabouts in order to be resistant to the following possibly true (but actually false) proposition: *Tom's twin was in the library when the book was stolen.*

The intuition embodied in this example is correct. If the twin really might have been in the library, S does not know that *Tom* is the thief. But my account of defective justification is able to capture that intuition without enlarging the class of genuine initiating defeaters beyond the set of actually true propositions. For the proposition that Tom has a twin who lives "in the vicinity" of the library is a true genuine initiating defeater. As discussed previously (in section 3.9), the boundary between knowledge and nonknowledge is fuzzy—one primary reason being that the extension of the class of genuine initiating defeaters is itself not clearly definable. But the important point to be noted here is that the set of possibly true (but actually false) propositions which the sceptic might wish to include among the class of defeaters will be related to some actually true propositions which are themselves genuine initiating defeaters. For, as in the Grabit Case just presented, whatever reason prompts the sceptic to include a possibly true, but actually false proposition among the class of defeaters will, itself, be a genuine initiating defeater. So-called "relevant alternatives" to a proposition, which must be eliminated by the evidence for the proposition in order for it to be absolutely certain, are relevant because of some actually existing set of facts.[61] No doubt, there will be cases in which it will not be clear whether the proposition representing those facts is a misleading or a genuine initiating defeater. But, presumably, that ambiguity will be shared by the corresponding, possibly true (but actually false) proposition which the sceptic wishes to place among those which can defeat the justification. Thus, the distinction between *certainty in the actual situation* (the situation in which Tom's twin is "in the vicinity") and *certainty in more situations than the actual situation* allows us to account for the varying intuitions about the extent of knowledge. Those divergent intuitions depend upon how large the boundaries of the "vicinity" are drawn (see section 3.9).

There is another, related reason for employing the distinction between a proposition's being certain in more situations than the actual one and a proposition's being more certain than another one in the actual situation. Recall Dretske's Zebra Case from Chapter Two. In that case, S was standing in front of a pen in the zoo. The sign above the pen read "Zebras," and S knew what zebras look like. Further, the animals in the pen looked like zebras. Thus, S has good evidence for the claim that all the animals in the pen were zebras. But consider the proposition that 'the animals in the pen are zebras or cleverly disguised mules'. It might be thought that this disjunctive proposition is more certain than the proposition 'all the animals in the pen are zebras'. After all, there are fewer *possible* defeaters of the disjunctive proposition than of the atomic proposition. For example, 'the animals are painted mules' is a defeater of the latter but not of the former.

But if we grant what the sceptic insists upon, and what I have accepted, namely, that 'certainty' is an absolute term which does not admit of degrees, then if 'the animals are zebras or painted mules' is more certain than 'the animals are zebras', the latter is not certain. That is merely an instantiation of EC2 (see sections 3.12 through 3.13). The sceptic would, no doubt, happily accept that conclusion. But now consider the proposition 'the animals are either zebras, painted mules, or cleverly disguised mechanical zebra-replicas'. Once again, if we fail to distinguish those propositions which are certain in more possible situations than others from those propositions which are more certain than others in the actual situation, we will be faced with the conclusion that our new proposition with three disjuncts is more certain than the one with only two disjuncts. And, once again, the sceptic may be pleased to note that 'the animals are either zebras or painted mules' is not certain. But this set of purportedly more and more certain propositions leads to the consequence already discussed. For the only justifications for propositions which would provide for certainty would be those justifications for which it is not *possible* that there are defeaters. And that has the consequence, as argued above, that the evidence for a proposition must entail it.

Nevertheless, there is the basic intuition, mentioned above, which the sceptic may still believe is not captured by my proposed analysis of certainty—the intuition that, if e makes p certain, the *truth* of p must be guaranteed by e. But note that, according to my proposal, it is not logically possible both that p is certain and that p is false, because if p is confirmed by e, and there is no genuine initiating defeater of that confirmation, then ∼p is not true. For ∼p would be such a defeater. Thus, if p is certain, p is true; and the first condition of knowledge, K1 (p is true), becomes redundant. In addition, if p is rendered certain by e, then e is true. For ∼e is a defeater of the confirmation of p by e. But it does not follow that e entails p. Even though it is not logically possible for p to be certain and false, it is not required that if p is certain, then it is not logically possible that p is false. For although, if p is certain on the basis of e, it is actually not the case that e is true and p is false, it is not further required that it is not possible that e is true and p is false. To interpret these requirements in that manner would be to commit a mistake of "misplaced necessity" similar to the one which I suggested that Almeder may have made in his criticism of the defeasibility theory of knowledge (see section 3.10).

Not only is the truth of p guaranteed by p's certainty in the manner just described, but, in addition, if e renders p certain, there is no other counterevidence which cannot be completely absorbed by e (see section 3.12). Hence, although a guarantee cannot be provided against all *possible* defeaters, it can be provided against all actual ones. *Merely* possible defeaters of a justification are no more detrimental than *merely* possible defects in a manufactured item.

Nevertheless, the sceptic may believe that a proposition cannot be rendered certain unless the evidence for it entails it. Recall the sceptic's rejoinder at the end of Chapter Two. There it was alleged that since knowledge entailed certainty, no acceptable account of knowledge was possible which allowed knowledge to be based upon defeasible justifications. In other words, the sceptic may continue to claim that if the evidence, e, for a proposition, p, is to *guarantee* p, e must entail p.

An aspect of the intuitive appeal of that claim is, no doubt, this: A feature of entailment is that if e entails p, then there is no proposition, i, such that (e&i) fails to entail p. The "guarantee" required by the sceptic is provided by the entailment because no proposition can block the entailment between e and p.

Note, however, that a very similar guarantee is provided by justifications which are in principle defeasible but in fact undefeated. For on my proposed account of certainty, if e renders p certain, then e confirms p *and* there is no non-misleading evidence, d, such that (d&e) fails to confirm p. In other words, if S knows that p on the basis of e, then p is guaranteed against all unabsorbable counterevidence. Of course, as mentioned above, this guarantee is provided against evidence and, consequently, it applies only to true propositions. But isn't this guarantee just as valuable in the actual situation as one which applies to the true and false propositions? Why would indemnification against all true and false propositions be more valuable in the actual situation? An insurance policy which provides coverage against all possible hazards is not worth more than one which provides coverage against all actual hazards.

Now, a reason which might be advanced for insisting on a guarantee against all propositions could be that if a proposition is to be certain, then if we know that it is certain, then we must know that there is no unabsorbable counterevidence. When we know that e entails p, then, other things being equal, we know that there is no proposition which can be conjoined with e to block the entailment to p. Whereas, when we know that e confirms but does not entail p, we do not know, other things being equal, that there is no nonmisleading evidence which can be conjoined with e to block the confirmation of p.

The reply to this reason for insisting on a guarantee against all propositions depends upon noting that my account of certainty is not such that a proposition is certain merely because it is confirmed, but rather it is certain because it is confirmed *and* there is no unabsorbable counterevidence. Thus, my proposed account of certainty does have the feature required. For if we know that e nondefectively confirms p, then we know, other things being equal, that there is no nonmisleading evidence which can block the confirmation.

I suspect, however, that what the sceptic may really mean to assert is something like this: If e entails p, then we can know quite easily that e renders p certain. All that is required is to recognize the entailment. Whereas, on my account, in order to know that e renders p certain, we must know that there are no nonmisleading defeaters. And that cannot be known simply by recognizing that e confirms p.

But we have already seen that this objection by the sceptic is beside the point. I am willing to grant that it is indeed more difficult to know that the conditions of knowledge and, hence, evidential certainty obtain than it is merely to have knowledge and hence, for a proposition to be evidentially certain. In the following sections, especially 3.16 and 4.1, I will present my reasons for thinking that propositions are *in fact* certain and that we can know that they are certain. Here the concern is whether my proposed account of certainty meets the requirements which can legitimately be imposed upon it by the sceptic and whether propositions *can* be certain. The issue is neither whether they are ever, in fact, certain nor whether we can know that they are certain. My claim here is merely that my proposal does capture important features of the intuition that evidence which renders a proposition certain guarantees the proposition. The fact that it is difficult to recognize that a proposition is certain does not prevent it from being certain.

I suspect, however, that the sceptic will be unhappy with this discussion of absolute certainty if only because the conclusion runs counter to the belief that contingent propositions can never be certain. In fact, many nonsceptics may be equally dissatisfied, for they, as well as the sceptics, have held that if knowledge entails absolute evidential certainty, then there is no knowledge of contingent propositions. The nonsceptics have rejected the entailment between knowledge and certainty, fearing that it leads to the impossibility of knowledge of contingent propositions. On the other hand, as mentioned above, the sceptic accepts the entailment and embraces the conclusion feared by the nonsceptic. I have disagreed with both and have argued that the entailment can be accepted without endangering our knowledge of contingent propositions. I hope that the nonsceptic will accept me as an ally, but I suspect that the sceptic will continue to search for some necessary feature of absolutely certain propositions, other than the ones considered thus far, which is such that it cannot be fulfilled by contingent propositions.

The sceptic may return to the considerations motivating the Evil Genius Argument in order to attempt to locate such an unfulfillable necessary condition of absolute certainty. The sceptic may claim that for all propositions (x, y) such that, if y is a contrary of x, or if y is a defeater of the justification of x, then, in order for x to be absolutely certain for S, ~y must be absolutely certain for S. This parallels the claim which I called the sceptic's Basic Epistemic Maxim and considered in detail in Chapter Two. That Basic Epistemic Maxim was: For all propositions (x,y) such that, if y is a contrary of x or if y is a defeater of the justification of x, then, in order for S to be justified in believing that x, S must be justified in believing that ~y. But, of course, once this parallel is noticed, my strategy for replying to this last attempt on the part of the sceptic to exclude contingent propositions from the class of absolutely certain ones should be apparent. For the arguments used in Chapter Two can simply be transferred here, with "absolute certainty" being substituted for "justification." I will not repeat that detailed argument. Suffice it to say that this principle is ambiguous in ways exactly parallel to the ambiguities in the sceptic's Basic Epistemic Maxim. Thus, in some senses, the principle is true, but it is useless to the sceptic for excluding contingent propositions from those which are absolutely certain. In other senses, the principle is false, but it would be useful if true. It is only the conflation of these interpretations which provides the sceptic with the hope that the consequences of the preceding discussion can be avoided.

In conclusion, it appears that EC2 (the requirement that there be no proposition more certain than p) is not only formally fulfilled by my proposal for absolute evidential certainty, but also that, properly understood, EC2 does not prevent an empirical, contingent proposition from being absolutely evidentially certain. Thus, I take it that we have removed the final objection to a proposition counting as evidentially certain when it is nondefectively justified. And, if that is correct, D2 has been fulfilled, since that desideratum required that we give an adequate account of absolute certainty.

treatment of this case is: What are the results of applying my partial analysis of justification, knowledge, and certainty to the Lottery Paradox, and are those results desirable?

So let us begin the examination of the Lottery Paradox. As Keith Lehrer points out, it is constructed to demonstrate that any theory of justification inevitably leads to paradoxical results which has as a feature that a proposition, say p, is adequately justified (for the sake of knowledge) by some evidence, e, even though e does not entail p.[63] Of course, if this were correct, a cornerstone of my proposed analysis of justification and my objections to Direct Scepticism will have been removed, and the entire structure will collapse.

Here is the Lottery Paradox as presented by Lehrer:

> The attempt to employ probability for guaranteeing the truth of non-basic beliefs reveals a crack in the structure of the foundation theory of justification that is quite beyond repair. Let us assume that the frequency probability statements are included among basic beliefs, and similarly that we include, among basic beliefs, assumptions to the effect that certain subjective and logical probability statements are reasonable estimates of frequencies. Such assumptions, rather than sustaining the foundation theory, provide for its destruction.
>
> We assume probabilities are ascertained. We cannot, however, equate high probability with complete justification. The set of statements that are highly probable on the basis of an evidence statement will be logically inconsistent with the evidence statement, and we cannot be completely justified in believing each of a set of inconsistent statements. Consider any set of statements $p1, p2, \ldots$ and so forth to pn, which describe outcomes of a lottery with one winning ticket and n consecutively numbered tickets in all. Statement $p1$ says the number one ticket will win in the drawing, $p2$ says that the number two ticket will win in the drawing, and so forth. Moreover, evidence statement e says that the drawing has been held and the winning ticket is one of the tickets numbered 1, 2, and so forth to n. Now consider the hypothesis $\sim p1, \sim p2$, and so forth to $\sim pn$, denying that the number one ticket, the number two ticket, and so forth is the winner. The $p(\sim p1,e) = 1 - 1/n$. Indeed, for any pj, $p(\sim pj,e) = 1 - 1/n$. We can imagine $1 - 1/n$ to be as large a fraction less than 1 than we wish by imagining n to be sufficiently large. So if one allows high probability to be any probability less than 1, it follows that each of the hypotheses $p1, p2$, and so forth to pn, is highly probable relative to e. The set of those hypotheses entails that none of the tickets numbered 1, 2, and so forth to n is a winner, in direct contradiction to the evidence

which asserts that one of those tickets is the winner. Thus, if statements having a probability less than one can be considered highly probable, and if highly probable statements are ones we are completely justified in believing, then we will be completely justified in believing a set of statements that are contradictory, guaranteeing that not all of them are true. This conclusion is entirely unacceptable.[64]

Since I will be referring frequently to the set of purportedly justified propositions, let us employ the following abbreviations, which avoid using the negation signs:

e_1 The probability of any p_i losing is $\frac{n-1}{n}$ ($1 \leqslant i \leqslant n$ and $n \geqslant 1$).

p_1 Ticket 1 will lose.

p_2 Ticket 2 will lose.

.
.
.
.
.
.

p_n Ticket n will lose.

The argument contained in the long quotation from Lehrer can be summarized as follows:

Suppose that S is justified in believing that p_1 on the basis of e_1 and that S is justified in believing that p_2 on the basis of e_1 . . . and that S is justified in believing that p_n on the basis of e_1. In our terminology, suppose that Jsp_1 & Jsp_2 . . . & Jsp_n. The proposition e_1 also justifies S in believing that not every ticket will lose; i.e., $Js{\sim}(p_1 \& p_2$ . . . $\& p_n)$. Hence, S is justified in believing each of two contradictory propositions: *all the tickets will lose* and *it is not the case that all the tickets will lose.* In our terminology:

$$Js(p_1 \& p_2 \; . \; . \; . \; \& p_n) \text{ and } Js{\sim}(p_1 \& p_2 \; . \; . \; . \; \& p_n).$$

I agree with Lehrer that any theory of justification is "entirely unacceptable" which has as a consequence that, for some x, both Jsx and Js~x. In fact, that would be a particularly ominous result for my proposed model of justification since, as I argued in section 2.8, one principle derivable from that model is the Contradictory Exclusion Principle, namely, that if S is justified in believing that x, then S is not justified in believing that ~x. Thus, if my analysis were committed to it being the case that for some x, Jsx and Js~x, then it would lead to the outright contradiction that Js~x and ~Js~x. If anything is an unacceptable consequence, surely that is!

It should be obvious that this is an important challenge to my analysis and to any analysis which permits a proposition to be completely justified on the basis of less than entailing evidence. Consequently, we must investigate the argument leading to the Lottery Paradox in order to determine whether it does have such fatal consequences. I hope to show that it does not.

The first point which should be noted is that Lehrer's statement of the argument leading to the paradoxical result contains an important ambiguity which tends to obscure an important feature of the argument. At one point, Lehrer says that "we cannot be completely justified in believing each of a set of inconsistent statements." To do so would be to be justified in believing each member of a set of propositions having a feature "guaranteeing that not all of them are true." Now, the set of propositions $\{p_1, p_2, \ldots p_n, \sim(p_1 \& p_2 \ldots \& p_n)\}$ is an inconsistent set of propositions, since they cannot all be true (together). But, at another point, Lehrer says that the "set of those hypotheses entails that none of the tickets numbered 1, 2 and so forth to n is a winner, *in direct contradiction* [emphasis added] to the evidence which asserts that one of those tickets is a winner." The set which he appears to have in mind here is $\{(p_1 \& p_2 \ldots \& p_n), \sim(p_1 \& p_2 \ldots \& p_n)\}$. Of course, this set—let us call it the *Lottery Paradox Set*—is also inconsistent. But there is a crucial difference. For only the Lottery Paradox Set contains a "direct contradiction."

It does seem that Lehrer has the Lottery Paradox Set in mind, because he refers to Henry Kyburg's treatment of the paradox in *Probability and the Logic of Rational Belief*. It is clear that Kyburg sees the paradox resulting from it being the case that S is justified in believing the "directly contradictory" propositions in the Lottery Paradox Set. He says, in the passage referred to by Lehrer:

> We *could* introduce a special convention to the effect that if A and B are ingredients of a rational corpus, then the conjunction of A and B is to be an ingredient of a rational corpus of this same level and basis. This has the effect of reinstituting the conventional logical rules concerning derivation from premises within the rational corpus of a given level. On the other hand, it would have seriously counterintuitive consequences. Consider, for

example, a lottery of a million tickets, of which one will be the winner. In advance of a fair drawing, the chances are a million to one against a given ticket being the winner. It seems reasonable then to include the statement, "ticket number j (for every j from 1 to 1,000,000) will not win the lottery," in a rational corpus of level r_i. If it is objected that this is not reasonable, on the grounds that there is a finite probability that ticket j will, after all, win the lottery, we can answer by pointing out that according to the same line of reasoning, there is a finite probability that *any* statistical hypothesis of the sort that everyone simply accepts, is false. But if we accept into the rational corpus of level r_i the statement (for every j) "Ticket number j will not win the lottery," and if we allow the conjunction of any two ingredients of a rational corpus to be also an ingredient of it, then the conjunction of the whole million of these statements will be an ingredient of our rational corpus, and this millionfold conjunction, together with the statement that there are only a million tickets, logically entails the statement, "None of the tickets will win the lottery." This statement, too, will therefore appear in a rational corpus of the given level and basis. Since we have in this rational corpus the statement that there is a ticket that will win the lottery, we may conjoin these two statements and derive any statement at all in the rational corpus of level r_i.[65]

(I should mention, parenthetically, that Kyburg goes on to say that it would be "intolerable" if we could derive any statement at all in the rational corpus of level r_i. He rejects what I will call step LP2 in the formulation of the paradox.).

Regardless of what Lehrer had in mind, it will be useful to examine the version of the paradox involving the Lottery Paradox Set, because it is usually understood in that manner.[66] Hence, I will assume that the version of the paradox under consideration depends upon showing that, on certain assumptions concerning justification, S is justified in believing a set of propositions, one member of which is a "direct contradiction" of another.

Thus, we will have to revise our formulation of the argument to make explicit a step contained in Lehrer's claim that if each of the propositions in the set $\{p_1, p_2, \ldots p_n\}$ is justified for S, then S is justified in believing that none of the tickets will win. Somewhere in the argument Lehrer must use a lemma to the effect that if (S is justified in believing that p_1 and S is justified in believing that p_2 . . . and S is justified in believing that p_n), then S is justified in believing that (p_1 and p_2 . . . and p_n). In our terminology, Lehrer must take the following to be true:

If $(Jsp_1 \, \& \, Jsp_2 \, \ldots \, \& \, Jsp_n)$, then $Js(p_1 \& p_2 \, \ldots \, \& p_n)$

Thus, the argument for the Lottery Paradox can be stated as follows:

LP1 e_1 is such that $Jsp_1 \, \& \, Jsp_2 \, \ldots \, \& \, Jsp_n$

LP2   If $(Jsp_1 \, \& \, Jsp_2 \, \ldots \, \& \, Jsp_n)$, then $Js(p_1 \& p_2 \, \ldots \, \& p_n)$

LP3 e_1 is such that $Js\sim(p_1 \& p_2 \, \ldots \, \& p_n)$

∴ $Js(p_1 \& p_2 \, \ldots \, \& p_n)$ and $Js\sim(p_1 \& p_2 \, \ldots \, \& p_n)$

The issue now becomes whether the argument is, in fact, sound. Briefly, my answer is that both LP1 and LP2 are false, so that, even though the argument is valid, it is not sound.

The reason that LP1 is false can be put simply: p_1 is not justified for S on the basis of e_1 because e_1 fails to confirm p_1. If this is correct, the same, of course, would hold for p_2, p_3, $\ldots p_n$. To see that this is correct, compare the evidence, e_1, for p_1 with the evidence, call it e_t, for p_1 given in what I will call the Testimonial Case: S is told by a reliable person, Mr. Testifier, that he (Mr. Testifier) saw some ticket other than number 1 drawn from the drum containing all the tickets in the lottery. So comes to believe e_t, where e_t is *Testifier, a reliable person, says that some ticket other than 1 was drawn.* Suppose that e_t is justified for S. S then concludes that p_1 on the basis of e_t.

I suggest that although, in both the Lottery Paradox Case (LPC) and the Testimonial Case (TC), the evidence for p_1 does not entail p_1, only in the TC is p_1 confirmed—where 'confirmed' is the constituent predicate in the justification model developed in section 2.7. In our terminology, $e_1 \, \mathcal{C} p_1$, but $e_t C p_1$. The crucial difference between e_1 and e_t is that only e_1 is "intrinsically" probabilistic. Evidence for p is *intrinsically probabilistic* if and only if it is composed only of a proposition which assigns *a high probability of less than one to p's truth* and a probability of greater than zero to ~p. The second clause in the definition is not strictly necessary, since it is entailed by the first clause. However, it is included to emphasize an important feature of intrinsically probabilistic evidence. The reason for requiring that *all* the evidence for p be probabilistic will emerge shortly. In the LPC, e_1 is such that the probability of any p_i losing is $\frac{n-1}{n}$ where $n \geqslant 1$. Hence, e_1 is intrinsically probabilistic.

It seems clear that, when the probability of p is sufficiently high, p becomes *acceptable as a hypothesis*. I agree with Henry Kyburg when he says, "my intuition is that no grounds can be offered for failing to accept the statement 'ticket 1 will not win' which cannot equally well be applied to many a scientific *hypothesis*. [Emphasis added.] "[67] Lehrer also refers to p_1 as a "hypothesis" in the passage quoted earlier.

Nevertheless, the most reasonable hypothesis may not yet be knowledge. Put another way, intrinsically probabilistic evidence can render a hypothesis acceptable but it cannot render a proposition confirmed. If the odds of ticket 1 winning are one in a million, it is certainly an acceptable hypothesis to adopt that ticket 1 will not win. But accepting a hypothesis that p or surmising that p on good grounds is not the same thing as acquiring knowledge that p. A hypothesis, even a very highly probable one, does not become knowledge until it is confirmed.

The primary reason for holding that, according to my proposal, evidence which is intrinsically probabilistic does not confirm p_1 (for the sake of knowledge) is that such evidence contains what I earlier called *unabsorbed* counterevidence for p_1 (see section 3.12). The evidence, e_1, assigns a high probability to p_1; but it also contains some evidence against p_1 —namely, that in a specifiable number of cases, ticket 1 will not lose. The high probability, $\frac{n-1}{n}$, that ticket 1 will lose does not absorb the counterevidence that there is a 1/n chance that ticket 1 will win. In fact, the evidence for p entails the unabsorbed counterevidence.

Now, it may be objected that I have misrepresented the evidence for p_1 because it is not necessary that e_1 contain intrinsically probabilistic evidence. It could be construed as merely being a proposition representing the lottery situation—i.e., there is a fair lottery with n tickets, etc. Hence, e_1, like e_t, can confirm p_1.[68]

There are two replies to this possible objection. First, presumably, the value of n is crucial for the confirmation of p_1. For if n equaled 2, for example, p_1 would not be confirmed. Hence, probabilistic evidence is, at least, implicit in e_1. Second, even if e_1 did not contain explicit mention of probabilities, it is the facts represented by e_1 which lead to the claim that ticket 1 has a very high probability, $\frac{n-1}{n}$, of losing. That claim, in turn, is offered as the confirming evidence for p_1. Consequently, although the anchoring link of the confirming evidence chain may not contain intrinsically probabilistic evidence, the link immediately preceding p_1 is intrinsically probabilistic. My comments can simply be transferred to the lack of a confirmation connection between that link and p_1.

Thus, I believe that LP1 is false and that we are not required to accept either of the two alternatives which Lehrer claims are forced upon us by the Lottery Paradox. He says of the argument leading to the paradox that it "shows that either we must deny that any statement with a probability of less than one is highly probable or we must deny that a highly probable statement is completely justified."[69] I believe that a proposition, p, may have a probability of less than one and still be completely justified, but not solely on the basis of evidence which is composed of *only* a proposition which explicitly assigns a high probability of less than one to p.

The evidence, e_1, fails to confirm p_1, not because e_1 fails to entail p_1, but rather because evidence which adequately confirms a proposition for the sake of knowledge must completely absorb any counterevidence which it contains. The evidence e_1 makes p_1 a reasonable surmise or hypothesis to adopt, but it does not confirm p_1, because p_1 is expected to be false on the basis of e_1 on some occasions when e_1 is true.

Before turning to my reasons for rejecting LP2, there is one other possible objection to my treatment of LP1 which should be considered, because it will help to clarify the distinction between evidence which is intrinsically probabilistic and evidence which is not intrinsically probabilistic. Suppose that someone were to claim the following: let us grant the distinction between evidence for p which is intrinsically probabilistic and evidence for p which is not intrinsically probabilistic. Furthermore, let us grant that intrinsically probabilistic evidence does not confirm p and that evidence which is not intrinsically probabilistic (but nevertheless not entailing) may confirm p. Now, reconsider the Testimonial Case (TC) in which $e_t C p_1$. Since e_t does not entail p_1, e_t can be true, and p_1 can be false. Presumably there have been some occasions when conditions similar to those represented by 'e_t' have obtained and conditions similar to those represented by 'p_1' have not.[70] Testimonial evidence is not always accurate. Thus, we can suppose that the probability of p_1 on the evidence e_t is $\frac{n-1}{n}$, where n is as large a number as the facts require. Perhaps such testimonial evidence is accurate 999 times out of 1,000 or 9,999 out of 10,000, etc. The probability can be made arbitrarily high by setting n at any chosen value. Let 'Q' be the true proposition containing the required value of n which represents the correct probability of p_1 on e_t.

The objection would continue by pointing out that it was granted that $e_1 \mathcal{C} p_1$ because e_1 was intrinsically probabilistic. But if $e_1 \mathcal{C} p_1$, it would appear that $(e_t \& Q) \mathcal{C} p_1$ since '$e_t \& Q$' contains a probabilistic assessment of p_1's truth on the evidence, e_t. If $(e_t \& Q) \mathcal{C} p_1$, then since Q is a true, nonmisleading defeater of the justification of p_1 by e_t, the proposition p_1 is not known by S according to my analysis of knowledge.

This is a perfectly generalizable result. For if the preceding objection were correct, every justification of a proposition, p, by evidence, e, would be defective if there is an occasion on which some conditions sufficiently similar to those represented by 'e' obtain and conditions sufficiently similar to those represented by 'p' do not obtain. For the probability of p on e would then be less than one. Thus, I may have exchanged the frying pan of the Lottery Paradox for the fire of scepticism.

The answer to this possible objection is that, although '$e_t \& Q$' contains a reference to the probability of the truth of p_1 on e_t, it is *not* intrinsically probabilistic evidence. Evidence for p_1 is intrinsically probabilistic when a probability of less than one of p_1's truth is the *only* evidence for its truth. The proposition '$e_t \& Q$' contains probabilistic evidence for p_1, but it *also* contains evidence, namely e_t, which is nonintrinsically probabilistic, and Q is completely absorbed by e_t.

To see that $(e_t \& Q)$ confirms p_1 consider a somewhat similar situation. Suppose that an epistemically reliable friend of S tosses a coin eight times and reports to S that it came up heads five times and tails three times. The objection will permit us to suppose that S can use that testimonial evidence, call it e_t, to confirm the claim that the coin did not come up heads on every toss. Now, if we conjoin to S's evidence the proposition, Q', that it is .996 probable that the coin will not come up heads on all eight tosses, the conjunctive proposition, $e'_t \& Q'$, does not diminish the confirmation of the proposition that it did not come up heads on all eight tosses. In the context of e'_t, the proposition, Q', does not count against the proposition that it did not come up heads on all eight tosses. In other words, Q' has been completely absorbed by e'_t.

Similarly, in the Testimonial Case, if we conjoin Q with e_t, the conjunctive proposition does not lessen the confirmation of p_1. Although we know that occasionally conditions similar to those represented by 'e_t' obtain and conditions similar to those represented by 'p_1' fail to obtain, that alone cannot serve to render the justification defective. For the possibility that e_t is true and p_1 false can be eliminated only if e_t entails p_1. The objection granted that e can confirm p even though e does not entail p. The proposition, p_1, is not supported *only* by evidence which assigns a high probability to p_1 — it is also supported by e_t, which confirms it, and Q is absorbed by e_t. Thus, this objection does not jeopardize the argument against LP1.

Let us turn to LP2. LP2 is an instance of a lemma which I called the converse of the Conjunction Principle, namely:

$$(x)(y)[\text{if Jsx and Jsy, then Js}(x\&y)]$$

In section 2.8, I gave my reasons for rejecting that lemma and perhaps I should reiterate them breifly. Recall that in order for a proposition, say x, to be justified for S, there must be a properly anchored proposition in S's reliably obtained belief set and a nondegenerate chain of propositions beginning with that properly anchored proposition and terminating in a proposition, say w, such that wCs *and* there is no grounded or pseudogrounded proposition, say u, for S such that $(w\&u)\mathcal{C}x$.

The lemma needed in the formulation of the argument for the Lottery Paradox is false, because there could be overriders of the conjunction, x&y, which are not overriders either of the confirmation of x or of the confirmation of y. For example, the proposition $\sim(x\&y)$ overrides any confirming evidence for $(x\&y)$ but it would neither override the confirming evidence for x nor the confirming evidence for y, unless either x confirmed y or y confirmed x. Consequently, S could be justified in believing that x and justified in believing that y but not justified in believing $(x\&y)$.[71]

In addition to the general reason for rejecting the lemma exemplified by LP2, there is a particular objection to its use in the Lottery Paradox Case because its employment violates the restrictions governing degenerate chains discussed in section 2.7. Recall that in order for a proposition to be justified for S it had to be confirmed by a proposition on a *nondegenerate* chain anchored in Γ_S^R. Presumably, the confirming evidence for the conjunction 'x&y' would include the confirming evidence for the conjunction 'x&y' would include the proposition 'x' and the proposition 'y', each located on separate but conjoinable confirmation chains whenever the converse of the Principle of Conjunction is employed as it is in LP2.

Thus, if for the sake of the argument we accept the converse of that principle, we must envision a series of confirmation chains joined at one link in the following pattern:

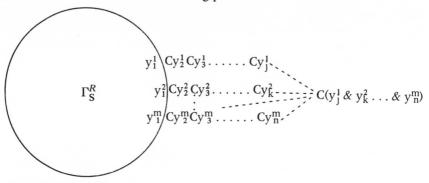

Figure 5.

Specifically, in the Lottery Paradox Case we must imagine that e_1 is an evidential ancestor of each of the conjunctions in the proposition $(p_1 \& p_2 \ldots \& P_n)$. Consequently, e_1 is an ancestor of the conjunction itself. In other words, in the Lottery Paradox Case the conjunction would be manufactured by fusing the n confirmation chains each of which contains e_1 in a link prior to the point at which the chain is fused. That particular structure could be represented as follows (assuming, contrary to fact, that LP1 is true):

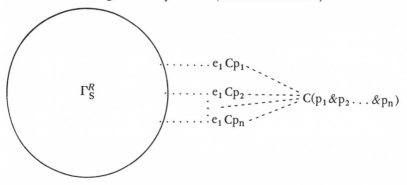

Figure 6.

It is crucial to recall that if the conjunction $(p_1 \& p_2 \ldots \& p_n)$ is to be justified for S, it must be located on a *nondegenerate* chain. We said in section 2.7 that a chain '$y_1 C y_2 C y_3 \ldots y_{n-1} C y_n$' degenerated at some link, say y_i, if y_i has an evidential ancestor, say y_j, which is such that $(y_j \& y_{i-1}) C y_i$. Thus, if the inference schema depicted by Figure 5 is a confirmation chain, the evidential ancestry of each conjunct in $(y_j^1 \& y_k^2 \ldots \& y_n^m)$ would have to be inspected in order to determine whether it contained a link which is an internal overrider of the confirmation of the conjunction by the set of conjuncts.

Now, it should be obvious that there is such an ancestral, internal overrider in the Lottery Paradox confirmation chain—namely, e_1. For every p_i is such that $(e_1 \& p_i) C (p_1 \& p_2 \ldots \& p_n)$ because e_1 confirms (if not entails) the denial of the conjunction, and this counterevidence, that is, e_1, is not absorbed at all by p_i. In other words, even if LP1 were true, and even if Figure 6 depicted a confirmation chain, the particular information chain in the Lottery Paradox Case terminating in '$(p_1 \& p_2 \ldots \& p_n)$' is a degenerate chain. Thus, $(p_1 \& p_2 \ldots \& p_n)$ would not be justified for S even if each conjunct in the compound proposition were justified for S and the converse of the Principle of Conjunction represented a generally acceptable inference pattern.

My objections to the Lottery Paradox can be summarized as follows: The argument underlying the paradox, though valid, is not sound because two of the premises are false. LP1 mistakenly asserts that intrinsically probabilistic evidence is sufficient to confirm a proposition for the sake of knowledge. LP2 is an instance of a false lemma. Consequently, the so-called Lottery Paradox presents no difficulties for my account of justification and cannot aid the sceptic.[72]

3.16 Some Contingent Beliefs Are Absolutely Certain

Thus far I have argued that the desiderata (D1 through D6) of an adequate account of absolute certainty, including all those which can reasonably be required by the sceptic, *can* be fulfilled by my proposed characterization of absolute evidential certainty. Contingent propositions *can* be absolutely certain on the basis of confirming evidence which does not entail them. Nothing prevents contingent propositions from being absolutely certain. In other words, contingent propositions *can* satisfy K3 and K4 in the analysis of knowledge developed and defended in sections 3.8 and 3.9.

But one final, important point needs to be made in this chapter. It is simply that we have good reasons for believing that the conditions of absolute evidential certainty *are* often actually fulfilled by contingent propositions. And, if that is correct, then we have good reasons for believing that S sometimes knows that p.

It is important to note that this final step in the argument of Chapter Three necessarily has a character different from that of the preceding ones in this chapter and in previous chapters as well. Up to this point we have been examining various epistemic concepts in order to determine whether the sceptic's claims about them were correct. Thus, for example, after examining the concept of justification and certainty, we saw that the sceptic had failed to provide any reason for believing that contingent propositions could be neither justified nor certain. In addition, we arrived at two general positive claims particularly relevant to this section:

(1) With regard to justification: S is justified in believing that p if and only if S has a grounded reason, w, which is adequate confirming evidence for p and there is no other grounded or pseudogrounded proposition, u, for S which overrides w for p.

(2) With regard to absolute evidential certainty: a proposition, p, is absolutely evidentially certain for S if and only if S is justified in believing that p on the basis of some evidence, w, and there is no genuine initiating defeater, d, of the justification of p by w.

As I mentioned above, we arrived at our rejection of the sceptic's arguments and at the characterizations of certainty and justification by a process of "conceptual analysis." However, the final step in the argument of this chapter cannot be established by further elucidation of concepts. At this point, all that I can do is to ask the reader to determine whether there are good reasons for believing that some of his/her beliefs or the beliefs of others actually fulfill the proposed conditions of justification and certainty. Thus, the major portion of this section will be devoted to clarifying the questions I want the reader to answer for himself/herself.

An analogy may help to show the difference between the task of this section and that of the previous ones. Suppose that, instead of considering the claims of the sceptics, we had been evaluating one of the forms of Zeno's paradoxical accounts of motion, say the claim that in a race the faster runner could not pass the slower one no matter how long the race, if the slower one had been given a head start. In addition, suppose that we had shown that one or more of the premises in the argument which led to the paradoxical conclusion were groundless and we have proposed some alternative, positive characterizations of the important concepts which were such that they both satisfied all the acceptable intuitions which informed the paradoxical account and *could*, in fact, be fulfilled. Thus, we would have shown that in a race between two runners, the faster one *could* win, provided that the racetrack was long enough. But we would not have shown that there were any good reasons to believe that a faster runner ever does pass a slower one. For by conceptual analysis we cannot show that the required conditions ever actually obtain. All races may, in fact, be too short, the faster runner may always stumble, or there may never actually be any races at all. That faster runners sometimes actually pass slower ones with head starts can only be "shown" by inspecting the actual world in order to determine whether we have good reasons for believing that there actually are cases of faster runners passing slower ones. Similarly, we cannot show by further conceptual analysis that some contingent propositions are absolutely certain. But I can ask the reader to ask the following two questions:

(1) Are there good reasons for believing that some contingent propositions are actually justified?

(2) Are there good reasons for believing that some contingent propositions are actually certain?

As I have already mentioned, most of this section will be devoted to clarifying these questions.

Let us begin with the question of justification. Since justification is a necessary prerequisite of certainty, if the answer to the first question is "no," there is no point in asking the second. We saw that a proposition, p, is justified for S if S has a grounded, nonoverridden, adequate reason, w, for p. So the first question resolves into three questions: Is there ever a proposition, w, which is grounded for S? Is w an adequate confirming reason for some contingent proposition, p? Is it ever the case that w is not overridden for p?

Let us begin with whether there is a proposition, w, which is grounded for S. Recall that, in order for a proposition to be grounded for S, it must either be in Γ_S^R—a subset of S's epistemically reliably obtained beliefs such that each member of Γ_S^R has no other reliably obtained belief in its evidential ancestry unless it is also in the evidential ancestry of the other beliefs—or attached to a proposition in Γ_S^R by a nondegenerate chain (see section 2.7). We can safely assume that *if* S has any reliably obtained beliefs, then at least one of them, call it 'w', is in Γ_S^R. For every reliably obtained belief is such that either it has another reliably obtained belief in its evidential ancestry or it does not have any beliefs in its ancestry. (Recall that if a nonreliably obtained belief is in the evidential ancestry of a belief, then that belief is not a reliably obtained belief.). If the reliably obtained belief has no other reliably obtained belief in its ancestry, it is in Γ_S^R; if it has a reliably obtained belief in its ancestry, then its ancestral chain either forms a circle or will terminate in an anchoring belief. If the former, all links of the chain will be members of Γ_S^R; if the latter, the anchoring link will be in Γ_S^R.

Thus, if S has any epistemically reliably obtained beliefs, some of them will be grounded for S. Presumably, however, the controversial aspect of the issue concerning the existence of grounded propositions is whether there are, in fact, any epistemically reliably obtained propositions. Let me postpone that issue for a moment.

Similarly, if the arguments in Chapters Two and Three have been sound up to this point, there should be no difficulty in answering the third part of the question concerning whether *if* w confirms p, it is ever nonoverridden. There need be no proposition, either grounded or pseudogrounded for S, which overrides the confirmation of p by w. For general sceptical reasons (the Evil Genius Argument, etc.) do not confirm any overriders, and we can suppose that S does not have any pseudogrounded (i.e., unconfirmed) propositions available as overriders, since we can give to S almost any set of beliefs we please.

Let me note in passing that I am not assuming that there is no defeater of the justification of p when I assume that, if S has adequate confirming evidence for p, it is not overridden. To say that a confirmation is *not overridden* is to restrict the counterpropositions to be considered to those propositions which are grounded or pseudogrounded for S. Those propositions may be true or false. On the other hand, to say that a *justification* is *not defeated* is to say, roughly, that there is no true proposition which is such that the conjunction of it with S's grounded or pseudogrounded propositions no longer provides adequate confirming evidence, unless it does so only because it renders a false proposition plausible. Thus, many nonoverridden justifications will be defeated, because there will be genuine initiating defeaters not grounded or pseudogrounded for S; but no nondefeated, justified belief can be overridden, because if a proposition is justified, it is already nonoverridden.

To return to the question of justification, given the assumptions just outlined, whether S is ever in fact justified in believing a contingent proposition depends upon whether S ever has any epistemicaly reliably obtained beliefs and whether any of those beliefs ever do in fact provide adequate confirmation of contingent propositions. Let me recall an example used in Chapter Two. There I suggested three sets of evidence, each of them purportedly adequate confirming evidence for the proposition that the table in my living room is brown. Those sets were:

e_1 The table looks brown to me, and nothing seems odd about the perceptual situation. Other people agree that it is brown.

e_2 Although I cannot now see the table, I bought a brown table this morning, and the furniture store where I purchased it confirms the delivery of the table. They said that it was placed next to a green chair in my living room. (I do have such a chair in my living room.)

e_3 If the table is brown, it will affect the spectroscope in some particular way, and it did.

Thus, the general question of whether we have good reasons for believing that S ever has adequate confirming evidence for a contingent proposition can be made more specific: Do we have good reasons for believing that e_1 through e_3 are ever reliably obtained, and, if so, are e_1 through e_3 adequate confirming reasons for believing that the table is brown? Before answering that more specific question, bear in mind that, as we have used the concept of adequate confirming evidence, e_1, for example, can be adequate confirming evidence for a proposition, but nevertheless it can be and often is overridden and/or defeated. In addition, recall that we have shown that principles like the following will not provide any basis for believing that e_1 through e_3 are not adequate confirming reasons:

MSEP   If S is justified in believing that p on the basis of e, then e must be adequate evidence for p; and the latter requires that e make p epistemically preferable to any contrary of p, say q.

I showed in Chapter Two that MSEP and other plausible sceptical epistemic principles posed no genuine obstacle to a proposition's being justified.

So, once again, let me ask: Do we have good reasons for believing that e_1, or e_2 or e_3 (or even their conjunction, for that matter) is ever reliably obtained and provides adequate confirming evidence for 'the table in my living room is brown'? Now, it may be thought that this question cannot be answered until a full characterization of epistemic reliability and adequate confirming evidence is provided, and it will, no doubt, be remembered that I eschewed such tasks in Chapter Two and Three. Although some of the features of both concepts have been discussed, I argued that such a full characterization was not necessary earlier; and it is not necessary here. For 'epistemic reliability' and 'adequate evidence' may be thought to be primarily descriptive notions which characterize the actual epistemological practices and general rules of the community — more binding and less variable than the rules of etiquette, but less precise and defined than the rules of chess. Or those concepts may be thought to be primarily normative notions which characterize a set of idealized epistemological rules and practices which have certain true-belief-producing characteristics such as consistency, coherence and congruence, testability, etc. If the former, then to say that e is reliably obtained and adequate confirming evidence for p is to say that the manner in which e was obtained and inferences to p from e are actually approved, sanctioned, or endorsed by a community of purported knowers;[73] if the latter, to say that e was acquired and is adequate confirming evidence for p is to say that the inference to p from e satisfies the requisite conditions of nondeductive reasoning.[74] Or, perhaps, those concepts are "mixed," such that both descriptive and normative characteristics play a fundamental, irreducible role.

I do not intend to speculate about that here, interesting as these issues are, because it is not necessary for the argument in this chapter. Moreoever, I believe that to do so would actually detract from the argument, for the answer to the question concerning the adequacy of e_1 through e_3 will be the same, whichever characterization of those concepts is adopted. It seems to be equally clear that e_1 through e_3 would in fact satisfy or fail to satisfy the demands of either the descriptive, normative, or mixed account of those epistemic concepts. Further, since I want as many readers as possible to agree with me that the conditions of absolute certainty are fulfilled, each reader is free to interpret those concepts in any manner which is consistent with what has been said about them in earlier sections.

I will return to the question concerning whether 'the table in my living room is brown' is justified in a moment, but I would now like to turn to the second general question of this section: Do we have good reasons for believing that 'the table in my living room is brown' is sometimes absolutely certain? Put another way: assuming for the moment that e_1 through e_3 are reliably obtained and do provide adequate confirming evidence for that proposition, are there good reasons for believing that the justification has no genuine initiating defeaters?

Now, of course, there *could* be genuine initiating defeaters of each subset. A defeater of e_1 *could* be: The perceptual situation is unreliable, because the light conditions are such that green tables often appear brown. A defeater of e_2 *could* be: The person making the delivery of the table is color-blind, and the packing carton was mislabeled. A defeater of e_3 *could* be: The spectroscope used to determine the color of the table was dropped shortly before it was used and it is no longer reliable. There *could* be such defeaters. But suppose that all of them were false and that there is no other genuine initiating defeater.

Isn't that sometimes the case? I think that the answer is clearly, "yes," but, as I have said, there is no *argument* which can be produced to show that it is. For the claim depends, in part, upon whether the table in my living room is, in fact, brown. For, if it were not brown, the proposition expressing the fact that it is not brown would be a genuine defeater. Just as I cannot show by an *argument* that the table is brown, I cannot show by an *argument* that there are no defeaters. But also, just as there *is* a way to show that the table is brown, there *is* a way to show that there are no genuine initiating defeaters. That way is, first, to explain carefully what is meant when someone asserts that the table is brown (if there is any doubt about what is meant) and, then, to point to the table and ask the reader to direct his/her attention to it (e_1); or call the furniture store where it was purchased to confirm that the brown table was delivered (e_2); or check the spectroscope (e_3); etc. Similarly, the way to show that there are no genuine intiating defeaters of the justification of the proposition that the table is brown is to explain carefully what is meant by a genuine initiating defeater and then ask the reader to ascertain whether there are any genuine initiating defeaters in the particular case at hand.

I think that I have explained the notion of 'genuine defeater' sufficiently, and if, after examination of the most plausible ones (specific ones like 'the lighting conditions in my living room are unusual' and general ones like those which motivate the Evil Genius Argument), the reader finds none, then he/she will have a good reason for believing that there are, in fact, none.

One more word of caution needs to be introduced before asking the general question, "Do we have good reasons for believing that contingent propositions are absolutely evidentially certain?" For in questioning whether we have good reasons for believing something, I am not asking whether we have conclusive reasons. In fact, to claim that we have good reasons to believe something is not even to claim that there are no good reasons, or even adequate ones, for that matter, against it. Of course, I have already argued in Chapter Two that there are no good reasons for the claim that S never knows that p, which implies that there are no good reasons for believing that contingent propositions are never evidentially certain. But here the question is the very modest one: Do we have good reasons for believing that some contingent propositions are absolutely evidentially certain?

Now, if the answer to that question is, "yes," then we would have a good reason for believing that on some occasions S knows that p. For if any one of S's beliefs, say p, is absolutely certain and believed (with psychological certainty) on the basis of the evidence which makes it so, then on that occasion S knows that p. To see that, recall the conditions of knowledge developed and defended in sections 3.8 and 3.9. They were:

S knows that p if and only if:

K1   p is true.

K2   S is certain that p on the basis of some proposition, e.

K3   e justifies p for S.

K4   every initiating defeater of the justification for S of p by e is a misleading initiating defeater.

We have already seen that S can be certain (psychologically) that p (see sections 2.11 and 3.4). Thus, if S's psychologically certain belief that p is absolutely evidentially certain on the basis of e:

K1 is fulfilled, because ~p would be a defeater of the justification of p (see section 3.4); and

K2 is fulfilled, because S is certain that p on the basis of e; and

K3 and K4 are fulfilled, because they jointly make up the necessary and sufficient conditions of absolute evidential certainty.

I have already argued that the particular defeasibility theory of knowledge developed in this chapter ought to be acceptable to the sceptics, that it squares with our intuitions in the clear as well as the vague cases of purported knowledge, and, finally, that it can withstand the attacks which have been developed specifically against it and defeasibility theories in general. In short, this analysis of knowledge and certainty seems correct. Thus, if we have good reasons for believing that some of S's beliefs are absolutely certain, then we have good reasons for believing that S sometimes knows that p.

So, for the last time: Do we have good reasons for believing that some contingent propositions are absolutely certain for S? As we have seen, that one question resolves into two component questions:

(1)   Do we have good reasons for believing that S sometimes has reliably obtained adequate confirming evidence for some contingent propositions?

(2)   Do we have good reasons for believing that some of the occasions when S does have reliably obtained, adequate confirming evidence for contingent propositions are also occasions when there are no genuine defeaters of S's justification?

I have tried to make (1) and (2) as clear as they can be. And I trust that the answers are also clear. I confess that I would not know how to respond to a person, who, understanding what I meant when I asserted that there was a brown table in my living room, when confronted with the brown table, claimed that he/she had no good reasons for believing that there was such a table. Perhaps all that could be done would be to begin again with the analysis of the sceptical principles examined in Chapter Two. Similarly, I believe that a person would either be feigning disbelief or entranced by scepticism who, understanding what I meant when I asserted that there are good reasons for believing that some contingent propositions are absolutely certain, claimed that he/she did not have those good reasons in situations in which he/she had evidence like e_1 through e_3 and no evidence for the existence of a genuine initiating defeater.

Now, if I am right that there are good reasons for believing that some contingent propositions are absolutely evidentially certain, then we are entitled to conclude that there are good reasons for believing that S sometimes knows that p.

CHAPTER 4

TYING IT ALL TOGETHER

4.1 Review of the General Goal and Strategy

This final chapter will be mercifully brief. For I merely wish to recall the goal and the general strategy of this book and show that we have, in fact, accomplished the task. Recall that the purpose was to show that three significant forms of scepticism are implausible. Those three varieties of scepticism are:

Direct Scepticism	It is not the case that S can know that p.
Iterative Scepticism	It is not the case that S can know that S knows that p.
Pyrrhonian Direct Scepticism	There are no better reasons for believing that S can know that p than there are for believing that it is not the case that S can know that p.

The notion of plausibility was defined in section 3.8 to mean that a proposition was plausible if and only if it is epistemically more reasonable to believe it rather than its negation. As I said when first introducing this definition, it is necessary to add the qualification "*epistemically* more reasonable" because there may be good reasons other than those having to do with the degree of evidence for a proposition which make it more reasonable to accept the proposition than deny it. Consequently, a proposition would have been shown to be implausible if a sound argument had been produced which demonstrates that it is more reasonable to believe the negation of the proposition than to believe it. Of course, some propositions would be neither plausible nor implausible if there were equally strong or weak reasons for believing it and its negation.

In order to refute a view, it seems to me that more is required than merely showing that it is not plausible; it must be shown to be implausible. Thus, for example, Direct Scepticism will have been refuted it if is epistemically more reasonable to believe that S can know that p than it is to believe that S cannot know that p. As I stated in Chapter One, I believe that all three forms of scepticism will have been refuted if the following four propositions can be shown to be true:

I   There is no good reason to believe that S never knows that p.
II   There are good reasons to believe that S sometimes knows that p.
III   There is a good argument for the claim that S sometimes knows that p.
IV   There are better reasons for believing that S sometimes knows that p than there are for believing that S never knows that p.

So two questions remain. Have I shown that I through IV are true? And, if I through IV are true, have all three forms of scepticism been refuted? Let us consider the latter question first, postponing the former for two subsequent short sections.

Direct Scepticism claims that S cannot know that p, where p is taken to be a contingent proposition normally believed to be known ('I have a pencil in my hand', for example). The reason given by the Direct Sceptic is that there is at least one necessary condition of knowledge which is such that, no matter how much evidence for p which S acquired, S's evidence would still fail to satisfy that condition. Now, if IV is true, Direct Scepticism is implausible, since the denial of IV is implied by the plausibility of Direct Scepticism. That is, Direct Scepticism asserts that S cannot know that p, and that, of course, implies that S never does know that p. Hence, if there are better reasons for believing that S sometimes knows that p than there are for believing that S never knows that p, then there are better reasons for believing that S can know that p than there are for believing that S cannot know that p.

Now, it might be objected that 'having better reasons' is not transmissible through entailment, just as some have held that 'justification' is not transmissible through deduction. There are two replies to this possible objection. First, the argument which I gave in Chapter Two (especially sections 2.5 and 2.8) against the view that justification is not transmissible through deduction is easily transferred here. Roughly, the argument would be that the better reason available for believing the consequent (as opposed to its denial) in the entailment is the antecedent itself. That is, if p entails q, and if S has better reasons for believing p than for believing its contradictory, then p provides S with a better reason for believing q rather than its (i.e., q's) contradictory. Second, as I also pointed out in section 2.4, the sceptic surely cannot object in this fashion, since the Evil Genius Argument—the core of the Direct Sceptic's position—depends upon the transmissibility of justification through entailment. That is, the Evil Genius Argument depends upon an instantiation of the general claim that justification is transmissible through entailment. Specifically, the argument depends upon the claim that if S is justified in believing that p, then S is justified in believing that there is no evil genius making S falsely believe that p. Thus, I believe that we can discard this objection; and that, in turn, secures our claim that, if IV is true, Direct Scepticism is implausible.

But IV is useful in another way. For if IV is true, Pyrrhonian Direct Scepticism is not only implausible, but it is false. Thus, if we have shown that IV is true, we have shown that Pyrrhonian Direct Scepticism is false, and that surely is sufficient to show that it is more reasonable to believe that it is false than to believe that it is true. Put another way: because Direct Scepticism is implausible, Pyrrhonian Direct Scepticism is false.

The iteration of the knowledge predicate in Iterative Scepticism makes it a more virulent form of scepticism—more immune to attack. For it may be that Direct Scepticism can be shown to be implausible and Pyrrhonian Direct Scepticism can be shown to be false; unfortunately, however, it does not follow directly that Iterative Scepticism is implausible, since it may be reasonable to believe that S does know that p while at the same time reasonable to believe that S does not know or, for that matter, cannot know that he/she knows that p. S may not know what the necessary conditions of knowledge are or that they obtain. Our task is to show that, if I through IV are true, then it is more reasonable to believe that S *can* know that he/she knows that p than it is to believe that S *cannot* know that he/she knows that p.

We must now ask: What are the conditions which are sufficient for S to know that he/she knows that p; and is it more reasonable to believe that all those conditions can obtain than it is to believe that one or more of them cannot obtain? The sufficient conditions of knowing that one knows ought to be merely an instantiation of the general conditions of knowledge; and, in order to remain faithful to the general strategy of always granting to the sceptic whatever can reasonably be granted, we ought to use the set of sufficient conditions of knowledge which have indeed been shown to be acceptable to the sceptic. Thus, we ought to use the analysis of knowledge developed and defended in Chapter Three (see especially sections 3.8 and 3.9).

That analysis was:

S knows that p if and only if:

K1   p is true.

K2   S is certain that p on the basis of some proposition, e.

K3   e justifies p for S.

K4   every initiating defeater of the justification for S of p by e is a misleading initiating defeater.

The issue, then, becomes: if I through IV are true, is it more reasonable to believe that all the following iterated conditions of knowledge (the sufficient conditions of S's knowing that *S knows that p*) can be fulfilled than it is to believe that they cannot be fulfilled?

IK1   *S knows that p.*

IK2   S believes that *S knows that p* on the basis of some proposition, e'.

IK3 e' justifies S in believing that *S knows that p.*

IK4   Every defeater of the justification for S of *S knows that p* on the basis of e' is a misleading defeater.

If I through IV are true, then it is clear that IK1 is more reasonable to believe than is its negation by IV. I take it that IK2 can be fulfilled, since it merely requires that S believes with certainty, on the basis of some proposition, e', that he/she knows that p; it does not require that e' justify that belief and it does not require that it is reasonable to believe that S knows that p on the basis of e'. I suppose that everyone will grant that S can believe with certainty almost anything on the basis of almost anything else.

The problematic conditions appear to be IK3 and IK4. That is, the issue seems to be whether there is an e' available to S which would justify the belief that *S knows that p* and, if so, whether there are any nonmisleading defeaters of the justification of *S knows that p* on the basis of e'.

The evidence that S *can* have is summarized by III. That is, the proposition that there is a good argument for the claim that S sometimes knows that p is the evidence, e', which can provide S with adequate evidence for the claim that S knows that p. Thus, if S has reviewed the reasons for and against the claim that S knows that p and has arrived at III through I and II, then S has evidence which justifies the belief that S sometimes knows that p. In other words, S can have a grounded, adequate confirming reason for believing that S sometimes knows that p. In short, S *can* have evidence, e', which justifies the belief that S sometimes knows that p.

Finally, it is more reasonable to believe that there are no genuine defeaters of the justification of the proposition that S knows that p on the basis of III than it is to believe that there are some genuine defeaters because I is true. For it asserts that there are in fact *no* reasons for believing that S does not sometimes know that p. Hence, there are no true propositions which would defeat the justification of the claim that S sometimes knows that p unless they rendered plausible the false proposition that there is some reason for believing that S never knows that p.

In sum, it is more reasonable to believe that each member of the set of propositions which are jointly sufficient for S's knowing that p can be fulfilled than it is to believe that one of them cannot be fulfilled. Now, it does not follow *directly* from the fact that it is more reasonable to believe that each of the four conditions can be fulfilled individually that it is more reasonable that they can *all* be fulfilled jointly. The so-called Lottery Paradox has taught us that it may be reasonable to believe each of a set of propositions individually and yet not be reasonable to believe the conjunction of the propositions (see section 3.15). But note that in order for Iterative Scepticism to be implausible, it is only required that we show that it is more reasonable to believe that S *can* know that S knows that p than it is to believe that S *cannot* know that he/she knows that p. But since nothing *prevents all four* of the jointly sufficient conditions of S's knowing that S sometimes knows that p from being fulfilled, if it is reasonable to believe that each is fulfilled in some circumstance (after S has read this book, for example), then it is reasonable to believe that they *can* all be fulfilled. That is, we have shown that S *could* actually bring it about that S knows that S knows that p. Hence, we have answered one of the questions reserved for this chapter: If I through IV are true, Iterative Scepticism as well as Direct Scepticism and Pyrrhonian Direct Scepticism will have been refuted.

Let me make a few more comments about the three forms of scepticism before considering the second question of this chapter. I have already mentioned an important difference between the refutation of Direct Scepticism and that of Pyrrhonian Direct Scepticism. We are in a position to show that Direct Scepticism is implausible and that Pyrrhonian Direct Scepticism is false. In fact, Direct Scepticism is implausible because Pyrrhonian Direct Scepticism is false. There is also an important difference between my proposed refutation of Direct Scepticism and the proposed refutation of Iterative Scepticism. For although both claim that S *cannot* know something ('p' or 'S knows that p'), we are in a position to show that it is more reasonable to believe that S does (in fact) sometimes know that p than it is to believe that S never knows that p; whereas we are only in a position to show that it is more reasonable to believe that S *can* know that S knows that p. But this is as it should be. For knowing that one knows that p is more difficult than knowing that p. The latter requires having adequate confirming evidence for p; the former requires having adequate confirming evidence for p and adequate confirming evidence for the claim that one has adequate confirming evidence for p. Thus, S may know that p without knowing that he/she knows that p. Having a good argument to show that S knows that p only provides a good reason for believing that S can bring it about that S knows that S knows that p—by acquiring the good argument. But if S is to acquire that argument, S must assess the adequacy of the evidence for p. That is, S must determine whether the sufficient conditions of knowledge that p obtain. Thus, in some sense only those who have examined the arguments for scepticism and found them to be inadequate know that they know that p.

To sum up the discussion thus far: We have answered affirmatively to one of the two questions set aside for this chapter. I have argued that if I through IV are true, Direct Scepticism, Pyrrhonian Direct

Scepticism, and Iterative Scepticism have been refuted—although the refutation of each takes a slightly different form. What remains is to briefly retrace the arguments for I and II. It is not necessary to present an independent argument for II or IV. IV merely summarizes I and II; and III refers to the argument which is the conjunction of I and II. That is, if I and II are true, both II and IV are true. So in the next two sections let me briefly restate the arguments for I and II.

4.2 There Is No Good Reason to Believe That S Never Knows That P

In Chapter Two, I examined the initially plausible reasons which the sceptic might advance for believing that S cannot (and, hence, does not ever) know that p. The Evil Genius Case provided the basis for the sceptic's claim that S cannot know that p, because if the sceptic's analysis of that case is correct, it shows that S cannot be justified in believing that p.

Roughly, my argument against the sceptic was that the Basic Epistemic Maxim underlying the Evil Genius Case was ambiguous in many ways and that, when the ambiguities were removed, the resulting interpretations of the maxim failed to provide any reason for believing that S never knows that p. That basic, ambiguous epistemic maxim was: In order for S to be justified in believing that p, S must be justifieid in rejecting the claim that there is some malevolent mechanism which is, or could be, making S falsely believe that p. I argued that every interpretation of that maxim was either such that it provided no reason for the claim that S cannot be justified in believing that p, or such that it contradicted the first of the two presuppositions underlying the dispute between the sceptic and the nonsceptic. There were two such presuppositions:

(1) If S knows that p on the basis of e, it is not required that e entail p; and

(2) If S knows that p on the basis of e, then e makes p absolutely evidentially certain.

As mentioned above, it is the first of these two presuppositions which created difficulties for the sceptic. But if (1) is not accepted as a presupposition of the dispute, the purported dispute vanishes.

For it would be readily granted, without argument, that the evidence which provides the basis for our purported empirical knowledge is not such that it entails those empirical propositions.

In short, I argued that some interpretations of the Basic Epistemic Maxim, even if true, would be just as useful (if not more so) for the nonsceptic as for the sceptic. Other interpretations violated the first presupposition. Since we discovered difficulties with all the initially plausible reasons for believing that S cannot know that p, then there is no good reason for believing that there are no occasions when S does know that p.

4.3 There Are Good Reasons to Believe That S Sometimes Knows That P

The argument in Chapter Three consisted of showing how the second presupposition, namely, that knowledge entails certainty, was *in fact* satisfiable while at the same time continuing to adhere to the restriction that e need not entail p. I argued that a proposition, p, could be rendered absolutely certain on the basis of another proposition, e, even though e did not entail p. That is, in spite of what some philosophers have believed, presupposition (2) can be satisfied without denying presupposition (1).

A particular version of the defeasibility theory of knowledge was suggested as the vehicle for having one's cake and eating it, too. Specifically, I argued that p was absolutely certain on the basis of e if e confirmed p and there were no genuine initiating defeaters of the justification of p by e. The strategy was to extract from the sceptic's position the necessary requirements of a proposition's being absolutely certain and then to show that all those requirements were fulfilled whenever a justification was *in fact* nondefective.

Once it was shown that, within this account of the defeasibility theory of knowledge, p could be absolutely certain on the basis of e, even though e did not entail p, the sceptic would have no reason for rejecting the defeasibility theory of knowledge as an adequate assessment of the necessary and sufficient conditions of knowledge—provided, of course, that there were no other valid general objections to the theory. The most initially plausible of those objections were considered and shown to be unfounded.

Thus, there are good reasons for believing that, on some occasions,

S knows that p, if there is a good reason for believing that, on some occasions, all the necessary and jointly sufficient conditions of knowledge as portrayed by the defeasibility theory are fulfilled. I argued that it was reasonable to believe that, on some occasions, they were fulfilled.

4.4 The Last Words

I believe that we have discovered a good remedy for scepticism. Hume thought that inattention, alone, could provide a remedy. Now, that may be a remedy of some sort, but it is surely not a philosophically satisfying one. And scepticism is, after all, a philosophical view supposedly arising after careful examination of the necessary conditions of knowledge. I think that I have shown that careful attention leads to a refutation of scepticism. But others will have to be the judge of that.

Notes

NOTES

Chapter One

1. David Hume, *Treatise of Human Nature*, ed. L. A. Selby-Bigge, 2nd ed. (Oxford: Clarendon Press, 1978), p. 218.

2. G. E. Moore, "Proof of the External World," *Philosophical Papers* (New York: Collier Books, 1959).

3. Roderick Chisholm, *Theory of Knowledge*, 2nd ed. (Englewood Cliffs, N.J.: Prentice Hall, 1977), p. 49.

4. See H. P. Grice, "Logic and Conversation," *The Logic of Grammar*, ed. Donald Davidson and Gilbert Harman (Encino, Calif.: Dickenson Publishing Co., 1975), pp. 64-153.

5. I use ⌜p⌝ to refer to the uttered sentence expressing p.

6. Chisholm, *Theory of Knowledge*, esp. Chap. 2, and Ludwig Wittgenstein, *On Certainty*, ed. G. E. M. Anscombe and G. H. von Wright (New York: Harper and Row, 1972), esp. pars. 246-254, 296, 376, 411, 429, 449, 467.

7. Other critics of scepticism have adopted the method I have chosen to remove scepticism's plausibility. But since the three forms of scepticism have not been clearly delineated, the attacks have often been directed at only one form.

8. Edmund Gettier, "Is Knowledge True Justified Belief?" *Analysis* 23 (1963): 121-123.

9. Thanks are due Robert Audi for suggesting this consideration.

10. Chisholm, "On the Nature of Empirical Evidence," in *Empirical Knowledge*, ed. Roderick M. Chisholm and Robert J. Swartz (Englewood Cliffs, N.J.: Prentice Hall, 1973), p. 232.

Chapter Two

1. René Descartes, *Meditations on First Philosophy*, ed. Elizabeth S. Haldane and G. R. T. Ross (Cambridge: Cambridge University Press, 1967), pp. 147-148.

2. Keith Lehrer, "Why Not Scepticism?" *Philosophical Forum* 11 (3) (1971): 283-298. Lehrer is apparently no longer convinced by this argument for scepticism. See his *Knowledge* (Oxford: Oxford University Press, 1974), esp. pp. 238-240.

3. Peter Unger, *Ignorance: A Case for Scepticism* (Oxford: Oxford University Press,

1975), esp. pp. 25, 108-109. In *Ignorance,* Unger repeats much of his argument first developed in "A Defense of Skepticism," *Philosophical Review* 80 (1971): 198-218. For a reply to that article, see James Cargile, "In Reply to 'A Defense of Scepticism'," *Philosophical Review* 81 (1972): 229-236; and Gerald Barnes, "Unger's Defesne of Scepticism," *Philosophical Studies* 24 (2) (1973): 119-124.

4. Lehrer, "Why Not Scepticism?" pp. 292-294.

5. Cicero *Academica* 2 (Lucullus). (Rackham translation.) 18. 56.

6. Cic. *Acad.* 2. 26. 84-85.

7. As Martin Bunzl has pointed out to me, there is one other possible interpretation of H; namely:

> H_3 e & there could be an evil genius (for all S knows) which is capable of bringing it about that S believes falsely that p.

But if this is a defeater, it could be handled in the same manner as H_2. If it is not a defeater, there is no evidential connection between it and p—thus, '$Jsp \rightarrow Js{\sim}H_3$' would be false.

8. Chisholm, *Theory and Knowledge,* pp. 41, 114.

9. Actually, Thalberg uses the name, "The Principle of the Transmissibility of Justification Through Deduction." Deduction is taken to be a process of inferring a proposition from one that entails it. Since I take 'S is justified in believing that x' not to imply 'S believes x', the issue does not depend upon whether S arrived at x, but rather upon whether S is entitled to arrive at x. I changed the name to be consistent with what I take to be the issue; namely, the transmissibility of justification through *entailment.*

10. For an argument designed to show that the Gettier issues are not crucial in epistemology, see Michael Williams, "Inference, Justification and the Analysis of Knowledge," *Journal of Philosophy* 75 (May 1978): 249-263.

11. Fred Dretske, "Epistemic Operators," *Journal of Philosophy* 67 (24) (December 1970): 1003-1013.

12. Irving Thalberg, "In Defense of Justified True Belief," *Journal of Philosophy* (66) (1969): 794-803; and "Is Justification Transmissible Through Deduction?" *Phil. Studies* 25 (1974): 357-364.

13. Dretske, "Epistemic Operators," pp. 1015-1016.

14. Thalberg, "Is Justification Transmissible" pp. 347-348.

15. Peter D. Klein, "Knowledge, Causality and Defeasibility," *Journal of Philosophy* 73 (November 1976): 792-812, esp. 806-808.

16. I am indebted to Robert Audi for pointing out this possible objection. I do not know whether he would accept by answer.

17. Grice, "Logic and Conversation."

18. Gail Stine, "Dretske On Knowing the Logical Consequences," *Journal of Philosophy* 68 (9) (1971): 296-299.

19. This example is a compilation of several cases presented by Jeffrey Olen in "Knowledge, Probability and Nomic Connections," *Southern Journal of Philosophy* 15 (Winter 1977): 521-525. David Shatz also presented a similar case in a paper, "Skepticism, Closure, and Reliability Theories of Knowledge," given at the APA, Eastern Division meetings, December 1979. I am not certain that Prof. Shatz would now think that this presents a difficulty for the Contrary Consequence Elimination Principle.

20. I. T. Oakley, "An Argument for Scepticism Concerning Justified Beliefs," *American Philosophical Quarterly* 13 (3) (1976): 221-228. He argues that scepticism follows from both accounts of justification as well as from an account of justification in which justifications can be infinitely long.

21. See my section 3.15 for a further discussion of the Lottery Paradox.

22. I am especially indebeted to Al Schepis for his help at this point in the argument. I had once mistakenly believed that SPG was correct, and he not only pointed out that it was inconsistent with what I had argued earlier, but he was kind enough to provide a substantial part of the corrected version of SPG.

23. Special caution must be taken when defining 'reliably obtained' conjunctions and disjunctions. As would be expected, if the chain is anchored by a conjunction, both conjuncts must be reliably obtained. But we cannot simply require that only one member of an anchoring disjunction be reliably obtained. For there may be a disjunction, say p v q, in which only q is reliably obtained. Suppose that ~q is also reliably obtained. Then p would be confirmed but not reliably obtained. I leave this problem for another occasion, since its solution does not affect the argument of this chapter.

24. I am indebted to Michael Smith for this possible objection. I do not know whether he would accept my answer.

25. Our notion of 'degenerate chains' will have to be expanded to include inspecting the evidential ancestry of overriders manufactured by conjoining links from several chains. For example, each conjunct in (t_1 will not win & t_2 will not win . . . & t_{100} will not win) has the same evidence anchoring its chain as the evidence which anchors the denial of the conjunction. Hence, the conjunction is not available as an overrider because it is the result of conjoining propositions anchored by evidence which overrides the conjunction. Thus, $\sim(t_1$ will not win & t_2 will not win . . . & t_{100} will not win) is justified for S and it is not overriden by the conjunction (t_1 will not win . . . & t_{100} will not win) (For a further discussion of this issue see section 3.15, esp. n. 72.)

26. Unger, *Ignorance*, pp. 24-25. Unger has Moore in mind as the opponent of scepticism in this passage. I deleted reference to Moore because the argument which I gave in the preceding section was different from Moore's. Moore claimed that he knew various propositions. I was merely arguing that the sceptic has failed to show that S does not know those propositions which we believe are knowable.

27. Ibid., p. 27.

28. Michael Smith develops a similar criticism of Unger's defense of scepticism in his article, "Unger's Neo-Classical Scepticism," *Mind* (forthcoming).

29. Such an argument appears to be endorsed by Unger in *Ignorance*, Chap. III (esp pp. 108ff). This repeats his argument presented in "In Defense of Skepticism." For another treatment of this issue, see Chisholm, *Theory of Knowledge*, pp. 116-118. There it is pointed out that Dewey held a similar view.

30. Under these circumstances, S would seem to have the "right to be sure" that p. See A. J. Ayer, *Problems of Knowledge* (Baltimore, Md.: Penguin Books, 1956), Chap One.

31. Several defeasibility theorists have claimed this. See, for example, Marshall Swain, "Epistemic Defeasibility," *American Philosophical Quarterly* XI (1) (January 1974): 15-25, esp. 22.

32. See Klein, "Knowledge, Causality and Defeasibility," p. 808.

33. The case was first introduced by Keith Lehrer and Thomas Paxson in "Knowledge: Undefeated Justified True Belief," *The Journal of Philosophy* 66 (8) (1969): 225-237, esp. 228.

34. Lehrer, "Why Not Scepticism?" pp. 293-294.

35. Suppose that it is claimed that this argument depends upon the false assumption that there is a true proposition other than ~p in every world in which ~p is true. That is, it *could* be claimed that there are some possible worlds in which no proposition other than p (or ~p) has a truth value. If there were such a world, then the argument designed to show that the Contrary Consequence Elimination Principle requires that e entails p would not be conclusive. But even if there is such a world, the proposition, 'q has no truth value in w_i', is

true. And once again there is a true contrary of p, namely, (~p & 'q' has no truth value in w$_j$). Hence, the denial of that contrary would be false, and the conjunction of the denials of all the contraries would be false. Hence, it is not possible for all the contraries to be true and p false.

Chapter Three

1. Ludwig Wittgenstein, *On Certainty*, ed. G. E. M. Anscombe and G. H. von Wright (New York: Harper and Row, 1972). The numbers in my text refer to paragraph numbers in *On Certainty*, assigned by the editors.

2. I refer the reader to an excellent treatment of Wittgenstein's *On Certainty*: Thomas Morawetz, *Wittgenstein and Knowledge* (Amherst: University of Massachusetts Press, 1978).

3. Ibid., p. 74.

4. See Wittgenstein, *On Certainty*, esp. pars. 462, 520-521, 487-489, 622. See also Morawetz, *Wittgenstein and Knowledge*, pp. 22-23.

5. A. J. Ayer, "Wittgenstein on Certainty," *Understanding Wittgenstein*, Royal Institute of Philosophy Lectures, vol. 7, 1972-1973 (New York: St. Martin's Press, 1974), p. 231.

6. This example is not original with me, but I cannot remember where I first encountered it. It has been used by many philosophers.

7. *Encylopedia of Philosophy*, 1st ed., s. v. "certainty."

8. Bertrand Russell, *Inquiry into Meaning and Truth* (London: George Allen and Unwin, 1940), pp. 17, 20.

9. Wittgenstein, *On Certainty*. It would be unfair not to mention that, in *On Certainty*, Wittgenstein objects to the tendency of philosophers to invent special linguistic practices. See esp. pars. 415 and 260. In the latter, he says, "I would like to reserve the expression 'I know' for cases in which it is used in normal linguistic exchange."

10. See Chisholm, *Theory of Knowledge*, Chap. One.

11. Ayer, "Wittgenstein on Certainty," p. 232.

12. Moore, *Philosophical Papers*, p. 222.

13. Idem.

14. Roderick Firth, "The Anatomy of Certainty," *Philosophical Review* 76 (1) (January 1967): 16. I have made minor notational changes in the quotations in order to make the passages more easily conform to the notation used throughout my essay.

15. Richard Miller, "Absolute Certainty," *Mind* 87 (January 1978): 62-64 passim.

16. Firth, "Anatomy of Certainty," pp. 17-18.

17. See Klein: "A Proposed Definition of Propositional Knowledge," *Journal of Philosophy* 68 (16) (1971): 471-482; "Knowledge, Causality and Defeasibility," *Journal of Philosophy* 73 (20) (1976): 792-812; "Misleading Evidence and the Restoration of Justification," *Philosophical Studies* 37 (1980): 81-89.

18. An interesting account of this is found in Harry Frankfurt, "Philosophical Certainty," *Philosophical Review* 71 (3) (1962): 303-327.

19. Ayer, *The Problem of Knowledge*, Chap. One.

20. Firth, "Anatomy of Certainty," p. 5.

21. This and the next three citations are from paragraphs of Wittgenstein's *On Certainty*.

22. Unger, *Ignorance*, p. 63.

23. Unger sometimes speaks of the contrapositive of (2); that is, if something is more flat (certain) than another thing, the latter is not really flat (certain). See Unger, *Ignorance*, pp. 66-68. But this is equivalent to (2).

24. Ibid., esp. pp. 66-68.

25. That Unger really intended the argument for scepticism to be concerned with psycho-

logical certainty (and not evidential certainty) can be seen in the following summary:

I have argued that, because we are certain of at most hardly anything, we know at most hardly anything to be so. My offering the argument I did will strike many philosophers as peculiar, even many who have some sympathy with scepticism. For it is natural to think that, except for the requirement of the truth of what is known, the requirement of 'attitude', in this case of personal certainty, is the *least* problematic requirement of knowing. Much more difficult to fulfill, one would think, would be requirements about one's justification, about one's grounds, and so on. And, in a way at least, I am inclined to agree with these thoughts. Why, then, have I chosen to defend scepticism by picking on what seems to be, excepting 'the requirement of truth', just about the easiest requirement of knowledge? [Ibid., p. 89.]

For a criticism of Unger's argument here which seems also to point to the conflation of what I have called 'psychological' and 'evidential' certainty, see James Cargile, "In Reply to 'A Defense of Skepticism'," *Philosophical Review* 81 (1972): 229-236.

26. Miller, "Absolute Certainty," p. 53.

27. Frankfurt, "Philosophical Certainty," p. 320.

28. Ayer, "Wittgenstein on Certainty," p. 232.

29. From now on I will use 'certainty' to mean absolute evidential certainty unless otherwise explicitly indicated.

30. Lehrer, *Knowledge*, pp. 338-339.

31. See n. 17. See also Steven R. Levy, "Misleading Defeaters," *Journal of Philosophy* 75 (12) (1978): 739-742; my reply, "Misleading 'Misleading Defeaters'," 76 (7) (1979): 382-386; and John Barker, "What You Don't Know Won't Hurt You?" *American Philosophical Quarterly* 13 (4) (1976): 303-308.

32. This raises an interesting possibility which need not be discussed here because, presumably, the sceptic will not be willing to accept it. It is this: Suppose that d_1 is such that there is an e_i and an e_{i+1} such that $(d_1 \& e_i)\mathcal{C}e_{i+1}$; but also suppose that S has a grounded proposition, say r_1, which is such that $(d_1 \& e_i \& r_1)\mathcal{C}e_{i+1}$. Would S know that h_1? S's evidence set is, so to speak, self-repairing, because it contains a restorer of the apparently defective evidence. We might want to say that, if r_1 were available in some *immediate* sense to S, then S does know that h_1 even though $(d \& e_i)\mathcal{C}e_{i+1}$. The sceptic would argue that, since *the* justification used by S to arrive at h_1 is defective, S does not know that h_1. There is something to that claim. In any case, since our policy is to grant whatever is plausible to the sceptic, we need not consider this possibility in any detail.

33. This notion is borrowed directly from Chisholm in his *Theory of Knowledge*. However, he calls it "having some presumption in its favor." See esp. pp. 8-10, 135 n.

34. Both my early proposal in "A Proposed Definition of Propositional Knowledge" and a similar and perhaps better-argued account by Risto Hilpinen in "Knowledge and Justification" (*Ajatus* 33 (1) (1971): 7-39) failed to note the existence of misleading initiating defeaters. These accounts were shown to be defective by Swain, "Epistemic Defeasibility"; Bredo Johnson, "Knowledge," *Philosophical Studies* 25 (4) (1974): 273-282; and Ernest Sosa, "Two Concepts of Knowledge," *Journal of Philosophy* 67 (3) (1970): 59-66.

35. Johnson, "Knowledge."

36. See D. M. Armstrong, *Belief, Truth and Knowledge* (Cambridge: Cambridge University Press, 1973), esp. Pt. III, p. 10, for a good defense of the claim that knowledge entails true belief.

37. I am indebted to David Shatz for pointing to the possible inability of my proposal to characterize noninferential knowledge. I do not know whether he would agree with my answer.

38. Barker, "What You Don't Know Won't Hurt You?" Much of what follows concerning Barker's proposal repeats what I said in "Misleading Evidence and the Restoration of Justification," *Phil. Studies* 37 (1980): 81-89.

39. Barker, "What You Don't Know Won't Hurt You?" p. 303.

40. Gettier, "Is Knowledge True Justified Belief?" pp. 121-123.

41. I am indebted to John Barker for suggesting this possible objection to me.

42. Barker, "What You Don't Know Won't Hurt You?" p. 306.

43. Gilbert Harman, "Knowledge, Inference and Explanation," *American Philosophical Quarterly* 5 (3) (1968): 172. The consideration of this case parallels my treatment of it in the paper, "Knowledge, Causality and Defeasibility," pp. 809-811.

44. The passage quoted from Harman is not altogether clear, but it appears that Harman is implicitly employing the false Defeater Consequence Elimination Principle when he says that Tom has no reason to believe that others do not share his (Tom's) belief that the Civil Rights Worker has been assassinated. Harman may be suggesting that, if Tom is justified in believing that the Civil Rights Worker has been assassinated, then he is justified in believing that other people do not have any evidence to the contrary. (See my section 2.12.)

45. See n. 31. Much of what I say here repeats what I have said in my reply, "Misleading 'Misleading Defeaters'."

46. Ibid., pp. 740-741.

47. I am indebted to correspondence with Lucey on this issue. He was reacting to an earlier version of my characterization of misleading defeaters, but his objection, if sound, would apply equally well to my present account.

48. Robert Almeder, "Defeasibility and Scepticism," *Australasian Journal of Philosophy* 51 (3) (1973): 238-244.

49. Lehrer, "A Fourth Condition of Knowledge: A Defense," *The Review of Metaphysics* 25 (1) (1970): 122-128; Klein, "Propositional Knowledge"; Marshall Swain, "Knowledge, Causality and Justification," *Journal of Philosophy* 69 (11) (1972): 291-300.

50. Almeder, "Defeasibility and Scepticism," p. 241.

51. Ibid., pp. 241-242.

52. Ibid., pp. 242-243.

53. This may not be true in every case, since e could be a conjunction of propositions, and it may be that some of the conjuncts of e could be false (thus making ~e true), and the remaining conjuncts could be sufficient to justify h. But it is sufficient for my purpose that there be some cases in which e is false and does defeat the justification; simply let the remaining conjuncts be insufficient to justify h.

54. Mary Gibson has suggested to me that Almeder may have implicitly assumed that the condition of nondefectiveness can guarantee that there are no unknowable defeaters only if we (those interested in assessing the extent of S's knowledge) could rule out the possibility of defeaters. We could do that, it might be thought, only if e entailed p. This is similar to his conflation of those occasions when S knows that p and when we could justifiably assert that S knows that p. It is not a necessary condition of S knowing that p that we be able to determine that S knows that p.

55. The apparent problem of circularity would have to be avoided by making the definition of knowledge recursive or by replacing "knowable" in the *definiens* with "believable and justifiable," since, if we believed a true justifiable proposition, d, which was such that d and e fail to confirm h, we would be prepared to "state specifically what would be enough" (as Austin requires) to produce a nondefective justification of h—one which was not defeated by d.

56. At this point the defeasibility analysis seems to share a remarkable structural feature in common with other recent nondefeasibility accounts. See, for example, Alvin Goldman, "Discrimination and Percpetual Knowledge," *Journal of Philosophy* 73 (20) (1976): 771-791.

57. In fact, it may even be the case that (e&c) is such that, in the context of e, c counts for p.

58. Unger, *Ignorance*, pp. 66-68.

59. In fact, these considerations seem to parallel those presented by Frankfurt in his account of relative psychological certainty. (See my section 3.5).)

60. For an interesting treatment of some related issues see James Bogen, "Wittgenstein and Skepticism," *Philosophical Review* 83 (3) (July 1974): 364-373.

61. See Goldman, "Discrimination"; Stine, "Dretske on Knowing"; Bogen, "Wittgenstein and Skepticism"; Shatz, "Skepticism."

62. See Henry E. Kyburg, *Probability and the Logic of Rational Belief* (Middletown, Conn.: Wesleyan University Press, 1961), p. 197n. Kyburg's more recent treatment of it can be found in his *Probability and Inductive Logic* (New York: Macmillan, 1970), p. 176 and Chap. 14.

63. Lehrer, *Knowledge*, p. 145.

64. Idem.

65. Kyburg, *Probability and the Logic of Rational Belief*, p. 197.

66. A. A. Derksen interprets the paradox in this fashion in his "The Alleged Lottery Paradox Resolved," *American Philosophical Quarterly* 15 (January 1978): 67-73. His article contains an excellent bibliography.

67. Kyburg, *Probability and Inductive Logic*, p. 192.

68. My thanks to Fred Schick for his suggestions at this point. I am not certain, however, that he would agree with my argument here.

69. Lehrer, *Knowledge*, p. 45.

70. I have used expressions like 'the probability of p_1 on e_t' rather unrestrictedly. What I have meant is: The probability of p_1 on e_t is the relative frequency of the occurrences of states of affairs sufficiently similar to those represented by 'p_1' with those states of affairs sufficiently similar to those represented by 'e_t'.

71. See Derksen, "Paradox Resolved," pp. 72-73. There he offers another reason for rejecting the principle that if (Jsx & Jsy), then Js(x&y).

72. I think that it should be mentioned that my treatment of justification does *not* prevent S from being justified in believing a set of inconsistent propositions. It does not involve what Lehrer called a "direct contradiction" and thus it does not have the disastrous consequence that, for some x, both Jsx and ~Jsx. S could be justified in believing such a set if S were justified in believing each member of a conjunction (individually) and also the denial of the conjunction. That could happen if every overrider "manufactured" by conjoining the conjuncts was produced by a degenerate chain. (See Chap. Two, n. 25.) Hence, although each conjunct would be grounded for S, the conjunction would not be grounded or even pseudogrounded for S. Consequently, S could be justified in believing each member of the following set of propositions: $\{p_1, p_2, p_3, \ldots p_n, \sim(p_1 \& p_2 \& p_3 \ldots \& p_n)\}$. We could block that result by relaxing even further the requirements for overriding propositions as the sceptic may wish. But aside from the superficial resemblance between this set of inconsistent justified beliefs and the Lottery Paradox Set, I can see no harmful consequences in granting that S can be justified in believing each member of this set. Of course, we must continue to insist that the lemma used in LP2 is false; for, if it were true, this set would become as objectionable as the Lottery Paradox Set.

73. This approach has recently been developed systematically by John Pollock in his book, *Knowledge and Justification* (Princeton, N.J.: Princeton University Press, 1974). See esp. Chap. II.

74. This approach has recently been developed systematically by Lehrer in *Knowledge*. See esp. Chaps. 7, 8, 9.

Glossary

GLOSSARY

A. Epistemic Terms

B_s — the set of propositions actually subscribed to by S—both occurrent and dispositional (52)

xCy — x confirms y (25)

$x\mathcal{C}y$ — x fails to confirm y (25)

xC^*y — x is in the confirmation ancestry of y (51)

$x\epsilon\Gamma_s =_{df}$ — $(x\epsilon B_s)\ \&\ \sim(\exists z)\ [(z\epsilon B_s)\ \&\ (zC^*x)\ \&\ (x\mathcal{C}^*z)]$

$Gsx =_{df}$ — $(x\epsilon\Gamma_s)\ v\ \big\{(\exists y_1,\dots,\exists y_n)\ [(y_1\epsilon\Gamma_s)\ \&(y_1Cy_2)\ \&\ (y_2Cy_3)\ \&\ \dots (y_{n-1}Cy_n)\ \&\ (y_n = x)]\ \&\ \sim(\exists y_i)\ (\exists y_j)\ (y_j\ \&\ y_{i-1}\mathcal{C}y_i)\big\}$

where:
$1 < i \leqslant n$
$j < (i-1) < i$

R_s — the set of propositions in B_s which are reliably obtained by S (59)

xRy — yC^*x and x's reliability depends essentially upon y (59)

$x\epsilon\Gamma_S^R =_{df}$ — $x\epsilon R_s\ \&\ \sim(\exists y)\ (y\epsilon R_s\ \&\ xRy\ \&\ y\mathcal{R}x)$ (59)

A_s — propositions in B_s that are not in Γ_s or Γ_S^R and which are subscribed to by S but not because S subscribes to any other proposition (68)

The number in parentheses refers to the page on which the explication of the term or principle listed in the Glossary can be found.

$Osux =_{df}$

(1) x is confirmed for S and u conjoined with an evidential ancestor, y_i, of x fails to confirm y_{i+1} *and* (2) u is a conjunction of propositions each of which is either in B_s or a link of a pure chain anchored by a conjunction of propositions each of which is in Γ_s or Γ_S^R or A_s. (69)

$Jsx =_{df}$

$(\exists w)[Gsw \ \& \ wCx \ \& \sim (\exists u)(Osux)]$ (70)

x is pseudogrounded for S

x is available for S to use as overriding evidence but not as confirming evidence (69)

E_s

the set of propositions which are either grounded or pseudogrounded for S (144)

xPy

x makes y epistemically more reasonable to believe than not y (145)

degenerate chains

the chain '$y_1 Cy_2 Cy_3 \ldots y_{n-1} Cy_n$' is degenerate at link y_i if and only if y_i has an evidential ancester, y_j, which is such that $(y_j \ \& \ y_{i-1})Cy_i$ (56)

pure chain

a chain which does not degenerate (56)

d_n is an *effective* defeater of the justification of h by e_n $\Big\} =_{df}$

$(d_n \ \& \ e_i)Ce_{i+1}$ (146)

d_1 is an *initiating* defeater of the justification of h by e_n for S $\Big\} =_{df}$

(1) d_1 is true and not a member of E_s and (2) there is a D-chain from d_1 to an effective defeater, d_n, of the justification of h by e_n for S. (146)

d_1 is a *misleading* initiating defeater of the justification of h by e_n for S $\Big\} =_{df}$

There is some false proposition, f, in *every* D-chain between d_1 and an effective defeater, d_n, of the justification of h by e_n for S and f occurs in a link in the D-chain prior to every link in which a false member of E_s occurs. (148)

S knows that p = df	K1 p is true
	K2 S is (psychologically) certain that p on the basis of some proposition, e.
	K3 e justifies p for S.
	K4 Every initiating defeater of the justification of p by e for S is a misleading initiating defeater. (150)
p is (evidentially) certain for S = df	K3 & K4 (150)

B. Epistemic Principles Accepted

Conjunctive Principle

If (x&y) is justified for S, then x is justified for S, and y is justified for S; but the converse is not true.   (45)

Contradictory Exclusion Principle

If S is justified in believing that x, then S is not justified in believing that ~ x.   (45)

Contrary Consequence Elimination Principle

For any propositions, x and y, (necessarily) if y is a contrary of x, then if S is justified in believing that x, then S is justified in believing that not y.   (24)

Disjunction Principle

If x is justified for S, or y is justified for S, then x v y is justified for S; but the converse is not true.   (45)

Nontransitivity Principle

If x justifies y for S, and y justifies z for S, it does not follow that x justifies z for S.   (45)

Transmissibility Principle

If x entails y, and x is justified for S, then y is justified for S.   (45)

C. Epistemic Principles Rejected

Contrary Prerequisite Elimination Principle

For any propositions, x and y, (necessarily) if y is a contrary of x, then if e is adequate evidence for x, then e contains not y.   (25)

Defeater Prerequisite Elimination Principle

For any propositions, x and y, (necessarily) if y is a defeater of the justification for x, then if e is adequate evidence for x, e contains not y.   (25)

Principle of the Transitivity of Confirmation	For all x and y and x, (necessarily) if xCy and yCz, then xCz. (32)
Partial Transitivity of Confirmation	For any propositions, x and y, and z, (necessarily) if x → y and zCx, then zCy. (31)
Simple Principle of Grounding	(x) (y) [(Gsx & xCy) → Gsy] (49)

D. Other Principles and Hypotheses

Sceptic's Basic Epistemic Maxim	S must be justified in rejecting H (H_c or H_d) in order to be justified in believing that p. (23)
Moderate Sceptical Epistemic Principle	If S is justified in believing that p on the basis of e [eJsp] , then e must be adequate confirming evidence for p; and the latter requires that e make p epistemically preferable to any contrary of p, say, q. (107)
Contraries Hypothesis (H_c)	c & ~ p & there is some mechanism, M, which brings it about that S believes (falsely) that p. (23)
Defeaters Hypothesis (H_d)	e & there is some mechanism, M, which could bring it about that S falsely believes that p without changing the truth of 'e'. (23)

Index

INDEX

Peter Klein was educated at Earlham College and at Yale University, where he received his Ph.D. in 1966. He is professor of philosophy at Rutgers University.